Genericity in
Polynomial
Optimization

Series on Optimization and Its Applications   ISSN: 2399-1593

(*formerly known as Imperial College Press Optimization Series* —
ISSN: 2041-1677)

Series Editor: Jean Bernard Lasserre *(LAAS-CNRS and Institute of Mathematics, University of Toulouse, France)*

Published

Series on Optimization and its Applications – Vol. 3

Genericity in Polynomial Optimization

Huy-Vui Hà

Vietnam Academy of Science and Technology

Tiến-Sơn Phạm

University of Dalat, Vietnam

W🌐 World Scientific

NEW JERSEY · LONDON · SINGAPORE · BEIJING · SHANGHAI · HONG KONG · TAIPEI · CHENNAI · TOKYO

Published by

World Scientific Publishing Europe Ltd.

57 Shelton Street, Covent Garden, London WC2H 9HE

Head office: 5 Toh Tuck Link, Singapore 596224

USA office: 27 Warren Street, Suite 401-402, Hackensack, NJ 07601

Library of Congress Cataloging-in-Publication Data

Names: Hà, Huy-Vui. | Phạm, Tiến-Sơn.

Title: Genericity in polynomial optimization / by Huy-Vui Hà (Vietnam Academy of Science and Technology), Tiến-Sơn Phạm (University of Dalat, Vietnam).

Description: New Jersey : World Scientific, 2017. | Series: Series on optimization and its applications ; volume 3 | Includes bibliographical references and index.

Identifiers: LCCN 2016046913 | ISBN 9781786342218 (hc : alk. paper)

Subjects: LCSH: Mathematical optimization. | Polynomials.

Classification: LCC QA402.5 .H295 2017 | DDC 519.6--dc23

LC record available at https://lccn.loc.gov/2016046913

British Library Cataloguing-in-Publication Data

A catalogue record for this book is available from the British Library.

Desk Editors: V. Vishnu Mohan/Mary Simpson

Typeset by Stallion Press

Email: enquiries@stallionpress.com

Printed in Singapore

Preface

The aim of this monograph is to present some basic aspects of genericity in polynomial optimization.

To explain roughly what we mean by genericity in this monograph, we would like to quote Askold Khovanskii, an outstanding specialist of singularity theory and algebraic geometry, who wrote[1] on "the ideology of general position".

"... Suppose that the outcome of some natural process, such as a physical experiment, is the graph of a function of one variable over a closed interval of finite length. One feels that such a function ought to have only finitely many roots. Can this be proved rigorously? "Classical mathematics" answers this question with an univocal "no": for any given closed subset of a closed interval (such as a Cantor perfect set) there exists an infinitely differentiable function whose set of zeroes is the given set. In the "mathematics of general position", however, the statement can be proved: The space of smooth functions over a closed interval contains a dense open set of functions which have finitely many zeroes, maxima, and minima. A "generic" function is a function from a dense open set. "Non-generic" functions form a hypersurface of codimension one in the space of functions. Various pathological effects, such as the existence of infinitely many

---

[1]Khovanskii AG. Newton polyhedra (Algebra and Geometry). In: Gindikin SG. (ed.) *Singularity theory and some problems of functional analysis.* Amer. Math. Soc. Transl. (2), Vol. 153, 1992; pp. 183–199.

zeroes over a closed interval of finite length, may of course occur, but for the overwhelming majority of functions no complications arise. It is natural to expect that a function occurring in some natural context (and not constructed specially as a counterexample) will be a function in general position.

For various classes of objects (functions, mappings, differential equations) one can often single out those in general position, which, first, constitute the overwhelming majority, and, second, behave in a much simpler way than an arbitrary object..."

In this monograph, by a polynomial optimization problem, we mean the following:

Given polynomials $f, g_i, h_j : \mathbb{R}^n \to \mathbb{R}, i = 1, \ldots, l, j = 1, \ldots, m$, compute

minimize $f(x)$
subject to $g_1(x) = 0, \ldots, g_l(x) = 0, \; h_1(x) \geq 0, \ldots, h_m(x) \geq 0.$

Inspired by the above ideology of general position (or, in other words, the ideology of genericity), we would like to define the class of generic polynomial optimization problems. This class must contain almost all of the problems and each problem in the class has "nice properties". With the help of some relatively recent developments in singularity theory and algebraic geometry, we can achieve this goal to a great extent.

To identify generic problems, we use the two important notions of Newton polyhedron and non-degeneracy conditions of a system of polynomials with respect to their Newton polyhedra.

Each set of polynomials is associated with some rational polytope in $\mathbb{R}^n$, called the Newton polyhedron of the system. The set of all polynomials (or systems of polynomials) with fixed Newton polyhedra forms a finite-dimensional space. In this space, a system is generic if it satisfies the so-called non-degenerate condition (with respect to the Newton polyhedron). The set of generic systems forms a dense and open subset. Moreover, this is an open set in the Zariski topology of the ambient space, or, equivalently, non generic systems are contained in the zero set of some polynomial.

It turns out that this non-degenerate condition is very convenient for controlling behavior at infinity of the gradient of a function that we need to minimize. This fact is very important when the feasible set of a problem is not compact. In particular, handling generic problems becomes much easier than handling non-generic problems.

In this monograph, the following "nice properties" of a generic problem are discussed:

(i) The restriction of the objective function on the constraint set is a Morse function, i.e. it has only non-degenerate critical points (consequently, these critical points are finite in number) and distinct critical values;

(ii) The objective function is coercive on the constraint set;

(iii) There is a unique optimal solution, lying on a unique active manifold, for which the strong second-order sufficient conditions, the quadratic growth condition and the global sharp minima hold;

(iv) Under local perturbations of the objective function, the set of active constraints is constant, and both the optimal solution and the optimal value function depend analytically on parameters. All minimizing sequences converge to the unique optimal solution;

(v) There exists a natural sequence of computationally feasible semidefinite programs whose solutions give rise to a sequence of points converging finitely to the optimal solution of the original problem;

(vi) There are global Hölderian error bounds for polynomial systems with exponents explicitly determined.

The monograph is organized as follows. Chapter 1 presents basic notions and some fundamental facts about semi-algebraic sets and maps which will be used in the next chapters. Most of the proofs will be given here with full details.

Chapter 2 introduces and presents some properties of sets of critical points and of tangency varieties of semi-algebraic functions. In particular, we show that the sets of critical values and of tangency values (at infinity) are finite.

Chapter 3 is on Hölderian error bounds, i.e. convenient functions that are surrogates for the distance of points in a test set to a given set. We present a systematic and unified treatment for the existence of global error bounds for closed semi-algebraic sets. In particular, we establish Hölderian error bounds for polynomial systems with exponents explicitly determined by the dimension of the underlying space and the number/degree of the involved polynomials.

Chapter 4 focuses on whether or not a polynomial optimization problem has an optimal solution. We establish the existence of optimal solutions for convex polynomial programs and for non-degenerate polynomial programs. It is worth emphasizing that generically, polynomial programs are non-degenerate.

Chapters 5 and 6 deal with well-posedness and genericity in polynomial optimization. It is shown that generically, polynomial optimization problems have very nice properties.

Chapters 7–9 address the problem of optimizing a polynomial function over a basic closed semi-algebraic set, which is NP-hard in general. The cases where the problem has or does not have an optimal solution are considered. By using sums of squares certificates for positive polynomials, we define a sequence of semi-definite programming relaxations, whose associated sequence of optimal values is monotone non-decreasing and converges to the optimal value of the original problem. We show that generically, finite convergence takes place.

The chapters of the monograph can (in part) also be read separately. Each chapter contains bibliographical notes.

The monograph is designed for a wide audience, in particular, graduate students and researchers. The readers are required to be familiar only with basic properties of polynomials, basic notions of topology and calculus of functions in real variables.

To keep the monograph reasonably short, we often have to refer to some results in semi-algebraic geometry and variational analysis without proof although complete references are given.

We have made an attempt to unify and relate many results in the literature. Some new results and new proofs are included. We do not develop the theory in its most general setting; our goal is to present the main problems, ideas and results in as natural a way as possible.

Chapter 1 was written by the first author and the other chapters were written by the second author. The general idea and structure of the monograph, as well as possible mistakes, belong to both of us.

Foreword

In Mathematics, powerful theorems usually come with a set of à priori conditions under which the results of the theorem are valid. An immediate, natural and important question is how restrictive are such conditions since they limit the applicability of the theorem. For instance, many optimization algorithms build up a sequence of iterates $(x_k, f(x_k))$, $k \in \mathbb{N}$, that in the limit as $k \to \infty$, should satisfy certain necessary optimality conditions (for instance, the celebrated Karush–Kuhn–Tucker (KKT) optimality conditions). Therefore, characterizing the class of optimization problems for which such optimality conditions are indeed necessary is very important. In particular, determining whether such problems are "generic" is an issue of primary importance for applications.

The authors of this monograph consider the above genericity issue in various questions related to polynomial optimization, i.e. optimization problems where the objective function as well as the functions that define the feasible set are all real multivariate polynomials. They define the class of "generic" polynomial optimization problems, i.e. a class of polynomial optimization problems with "nice properties" (in terms of the objective function and the minimizers) and which contains "almost all" problems.

The distinguishing feature of the book by Huy-Vui Hà and Tiến-Sơn Phạm is to show that when restricting to polynomials, this class of generic problems can be defined in a simple and elegant manner by using only the two basic (and relatively simple) notions of

Newton polyhedron and non-degeneracy conditions (with respect to the Newton polyhedron) associated with a given polynomial optimization problem. Hence the semi-algebraic framework (which is still a very general setting encompassing a multitude of important applications) allows a (relatively) simple and unified treatment of genericity issues. As a consequence, the required background is kept to its minimum and makes this book an essential complement to existing books presenting the theory and algorithms of polynomial optimization.

Jean Bernard Lasserre

Acknowledgments

We wish to express our gratitude to Jean Bernard Lasserre for his suggestions and encouragements throughout the writing of this monograph. We also like to thank Tạ Lê Lợi for his useful comments and corrections.

Contents

Semi-algebraic Geometry

Abstract

Semi-algebraic sets and maps are objects, which, so to speak, are very "tame". (Hence, it is rather easy to work with them.) We will see, among the other things, that:

(i) The family of semi-algebraic sets is closed under taking finite unions, finite intersections, the closure and the complements. The image of a semi-algebraic set by a projection is also semi-algebraic.

(ii) Tame structure: Each semi-algebraic set can be decomposed into a finite disjoint union of the so-called cells, which are semi-algebraic subsets of very simple structure.

(iii) Topologically, semi-algebraic sets are tame: Every semi-algebraic set has only a finite number of connected components and each component is also semi-algebraic. A connected semi-algebraic set is locally connected and arcwise connected.

(iv) Analytically, semi-algebraic functions behave very tamely: For an arbitrary semi-algebraic map, we can decompose its domain of definition into a finite number of cells such that the restriction of the map on each cell is analytic. For semi-algebraic functions, the Łojasiewicz inequality and gradient inequality hold true. Also, there is a strong version of the Sard theorem.

1.1 Algebraic and semi-algebraic sets

Definition 1.1. A set of the form

$$\{x \in \mathbb{R}^n \mid f_j(x) = 0, j = 1, \ldots, k\},$$

where all f_j are polynomials, is called an *algebraic subset* of $\mathbb{R}^n$.

1

We remark that

$$\{x \in \mathbb{R}^n \mid f_j(x) = 0, j = 1, \ldots, k\} = \{x \in \mathbb{R}^n \mid f(x) = 0\},$$

where $f(x) := \sum_{i=1}^{k} f_j^2(x)$, which is a polynomial.

It is evident that the class of algebraic subsets of $\mathbb{R}^n$ is closed under taking finite unions, finite intersections, but not closed under taking complement.

Definition 1.2. (i) A subset of the form

$$\{x \in \mathbb{R}^n \mid f(x) = 0, f_j(x) > 0, j = 1, \ldots, k\},$$

where all f and f_j are polynomials, is called a *basic semi-algebraic set* of $\mathbb{R}^n$.

(ii) A subset of $\mathbb{R}^n$ is called *semi-algebraic* if it is a finite union of basic semi-algebraic subsets.

Clearly, each set of the form $\{x \in \mathbb{R}^n \mid f(x) = 0, f_j(x) \geq 0, j = 1, \ldots, k\}$ is semi-algebraic and we call it a *basic closed semi-algebraic set*.

Proposition 1.1. *Let A and B be semi-algebraic subsets of $\mathbb{R}^n$. Then the sets $A \cup B, A \cap B$ and $A^c := \mathbb{R}^n \backslash A$ are also semi-algebraic.*

Proof. It follows directly from definition that if A and B are semi-algebraic subsets, then $A \cup B$ and $A \cap B$ are also semi-algebraic.

Now, assume that $A = \bigcup_{i=1}^{s} A_i$, where A_i is a basic semi-algebraic subset of $\mathbb{R}^n$. We will verify that the complement A^c of A is a semi-algebraic subset of $\mathbb{R}^n$. It is enough to consider the case $s = 2$. Let

$$A_i = \{x \in \mathbb{R}^n \mid f_i(x) = 0, f_{i_j}(x) > 0, j = 1, \ldots, k_i\},$$

for $i = 1, 2$ and $A = A_1 \cup A_2$. Then

$$A_i = \{x \in \mathbb{R}^n \mid f_i(x) = 0\} \bigcap_{j=1}^{k_i} \{x \in \mathbb{R}^n \mid f_{i_j}(x) > 0\}.$$

Hence,

$$A_i^c = \{x \in \mathbb{R}^n \mid f_i(x) = 0\}^c \bigcup_{j=1}^{k_i} \{x \in \mathbb{R}^n \mid f_{i_j}(x) > 0\}^c.$$

Clearly, we have

$$\{x \in \mathbb{R}^n \,|\, f_i(x) = 0\}^c$$
$$= \{x \in \mathbb{R}^n \,|\, -f_i(x) > 0\} \cup \{x \in \mathbb{R}^n \,|\, f_i(x) > 0\}$$

and

$$\{x \in \mathbb{R}^n \,|\, f_{i_j}(x) > 0\}^c$$
$$= \{x \in \mathbb{R}^n \,|\, -f_{i_j}(x) > 0\} \cup \{x \in \mathbb{R}^n \,|\, f_{i_j}(x) = 0\}.$$

Therefore, the sets A_1^c and A_2^c can be written in the form

$$A_1^c = C_1 \cup \cdots \cup C_{s_1},$$

and

$$A_2^c = D_1 \cup \cdots \cup D_{s_2},$$

where each of the sets C_i, D_j is of the form $\{x \in \mathbb{R}^n \,|\, g(x) = 0\}$ or $\{x \in \mathbb{R}^n \,|\, g(x) > 0\}$ with g being a polynomial. Hence, each $C_i \cap D_j (i = 1, \ldots, s_1, j = 1, \ldots, s_2)$ is a basic semi-algebraic set.

Now, since

$$A^c = (A_1 \cup A_2)^c$$
$$= A_1^c \cap A_2^c = \left(\bigcup_{i=1}^{s_1} C_i \right) \cap \left(\bigcup_{j=1}^{s_2} D_j \right) = \bigcup_{i=1}^{s_1} \bigcup_{j=1}^{s_2} C_i \cap D_j$$

is a finite union of basic semi-algebraic sets, it is semi-algebraic. $\square$

Remark 1.1. The class $\mathcal{SA}_n$ of semi-algebraic sets in $\mathbb{R}^n$ is the smallest class of subsets of $\mathbb{R}^n$ such that

(i) If P is a polynomial, then

$$\{x \in \mathbb{R}^n \,|\, P(x) = 0\} \in \mathcal{SA}_n \quad \text{and} \quad \{x \in \mathbb{R}^n \,|\, P(x) > 0\} \in \mathcal{SA}_n.$$

(ii) If $A \in \mathcal{SA}_n$ and $B \in \mathcal{SA}_n$, then $A \cup B, A \cap B$ and A^c are in $\mathcal{SA}_n$.

Example 1.1. (i) A semi-algebraic set in $\mathbb{R}$ is a finite union of points and open intervals.

(ii) Every algebraic set is semi-algebraic.

(iii) Let $\mathcal{P}_n$ be the subset of all positive definite symmetric matrices in the space $\mathcal{M}_n$ of all $n \times n$ matrices. Identifying $\mathcal{M}_n$ with $\mathbb{R}^{n^2}$. Then $\mathcal{P}_n$ is defined by

$$\mathcal{P}_n = \left\{ (a_{ij}) \in \mathbb{R}^{n^2} \,\middle|\, \sum_{i,j=1}^{n} (a_{ij} - a_{ji})^2 = 0, f_k((a_{ij})) > 0, k = 1, \ldots, n \right\},$$

where f_k is the kth principal minor[1] of the matrix (a_{ij}), and hence $\mathcal{P}_n$ is a basic semi-algebraic subset of $\mathbb{R}^{n^2}$.

(iv) The following sets are not semi-algebraic:

$$\{(x,y) \in \mathbb{R}^2 \,|\, y - \sin x = 0\}, \quad \{(x,y) \in \mathbb{R}^2 \,|\, y - e^x = 0\},$$

$$\{(x,y) \in \mathbb{R}^2 \,|\, y = nx, n \in \mathbb{N}\}.$$

1.2   Some lemmas

The following simple lemmas are at the heart of many arguments concerning semi-algebraic sets and functions.

We associate a point $a := (a_0, \ldots, a_d) \in \mathbb{R}^{d+1}$ with the polynomial

$$P_a(t) := a_0 + a_1 t + \cdots + a_d t^d$$

in one variable t. For integers k and l with $1 \leq k \leq l \leq d$, we put

$$\mathcal{P}_k^l := \{a \in \mathbb{R}^{d+1} \,|\, \deg P_a = l \text{ and } P_a \text{ has exactly } k \text{ distinct}$$
$$\text{complex roots}\}.$$

Lemma 1.1. $\mathcal{P}_k^l$ *is a semi-algebraic set.*

Proof. We establish two claims.

---

[1]kth principal minor of a square matrix A is the determinant of its submatrix formed by the intersection of the first k rows and the first k columns of A.

Claim 1: *Let $f(t) := a_0 + \cdots + a_l t^l$, $a_l \neq 0$, and $k \leq l$. Then f has at most k distinct complex roots if and only if there exist polynomials*

$$r_\xi(t) = \xi_0 + \xi_1 t + \cdots + \xi_{k-1} t^{k-1}$$

and

$$q_\xi(t) = \xi_k + \cdots + \xi_{2k} t^k$$

for some $\xi := (\xi_0, \ldots, \xi_{2k}) \in \mathbb{R}^{2k+1} \setminus \{0\}$ such that

$$f(t) r_\xi(t) = f'(t) q_\xi(t), \tag{1.1}$$

where f' is the derivative of f.

Proof. Let $D \in \mathbb{R}[t]$ be the greatest common divisor of the polynomial f and f'. Then there are polynomials $q, r \in \mathbb{R}[t]$ such that $f(t) = D(t) q(t)$ and $f'(t) = D(t) r(t)$. Hence,

$$f(t) r(t) = f'(t) q(t).$$

Furthermore, $\deg q = l - s$ and $\deg r = l - s - 1$, where $s := \deg D$. The number of distinct complex roots of $f(t)$ is equal to $l - s$. Hence, if $k \geq l - s$, then (1.1) will hold with $r_\xi(t) = r(t)$ and $q_\xi(t) = q(t)$.

Conversely, assume that there exist polynomials

$$r_\xi(t) := \xi_0 + \cdots + \xi_{k-1} t^{k-1}$$

and

$$q_\xi(t) := \xi_k + \cdots + \xi_{2k} t^k$$

with some $\xi = (\xi_0, \ldots, \xi_{2k}) \in \mathbb{R}^{2k+1} \setminus \{0\}$ such that (1.1) holds, then $f r_\xi$ is a common multiple of f and f'. Hence, the degree of the polynomial $f r_\xi$ must be greater or equal to the degree of the least common multiple of f and f':

$$\deg(f r_\xi) \geq 2l - s - 1.$$

Note that $l + k - 1 \geq \deg f r_\xi$. Therefore, $k \geq l - s$, which means that f has at most k distinct complex roots. $\qquad\square$

Claim 2: *For $1 \leq k \leq l$, the set*

$$\mathcal{Q}^l_{\leq k} := \{a = (a_0, \dots, a_l) \in \mathbb{R}^{l+1} \,|\, a_l \neq 0 \text{ and}$$
$$P_a(t) \text{ has at most } k \text{ distinct complex roots}\}$$

is a semi-algebraic subset.

Proof. By Claim 1, for each $a \in \mathbb{R}^{l+1}$ with $a_l \neq 0$, the polynomial $t \mapsto P_a(t)$ has at most k distinct complex roots if and only if (1.1) holds for

$$r_\xi(t) = \xi_0 + \cdots + \xi_{k-1} t^{k-1}$$

and

$$q_\xi(t) = \xi_k + \cdots + \xi_{2k} t^k$$

with some $\xi \in \mathbb{R}^{2k+1} \backslash \{0\}$. Equating the coefficients of t^j, $j = 0, \dots, l+k-1$ in (1.1), we obtain a system of $l+k$ homogeneous linear equations

$$\beta_0(a; \xi_0, \dots, \xi_{2k}) = 0,$$
$$\vdots$$
$$\beta_{l+k-1}(a; \xi_0, \dots, \xi_{2k}) = 0,$$

of $2k+1$ unknowns $\xi_0, \dots, \xi_{2k}$.

Let $k \leq l-1$. Then the number of unknowns is at most equal to the number of equations. Hence, the system has non-zero solution ξ if and only if all $(2k+1)$-minors of the matrix of the system are vanished. Since each of the minors is a polynomial in $(a_0, \dots, a_l)$, the set $\mathcal{Q}^l_{\leq k}$ is semi-algebraic.

Assume that $k = l$. Then each root of $P_a(t)$ is simple. Hence,

$$\mathcal{Q}^l_{\leq k} := \{a \in \mathbb{R}^{l+1} \,|\, \Delta(a) \neq 0\},$$

where $\Delta(a)$ is the discriminant of $P_a(t)$. Since $\Delta(a)$ is a polynomial in the variable a (with coefficients in $\mathbb{Z}$), the claim is proved. $\qquad\square$

Now, we are ready to finish the proof of Lemma 1.1. Indeed, we see that

$$\mathcal{P}_k^l = \{a \in \mathbb{R}^{d+1} \mid \deg P_a = l \text{ and } P_a \text{ has exactly } k \text{ distinct}$$
$$\text{complex roots}\}$$
$$= \{a \in \mathbb{R}^{d+1} \mid a_d = \cdots = a_{l+1} = 0 \text{ and } (a_0, \ldots, a_l) \in \mathcal{Q}_{\leq k}^l\}$$
$$\setminus \{a \in \mathbb{R}^{d+1} \mid a_d = \cdots = a_{l+1} = 0 \text{ and } (a_0, \ldots, a_l) \in \mathcal{Q}_{\leq k-1}^l\}.$$

It follows from Claim 2 that $\mathcal{P}_k^l$ is a semi-algebraic set of $\mathbb{R}^{d+1}$. □

Definition 1.3. Let $A \subset \mathbb{R}^n$ and $f \colon A \to \mathbb{R}$. The function f is said to be *analytic at a point* $p \in A$ if there exists an open neighborhood U of p in $\mathbb{R}^n$ and an analytic map $\hat{f} \colon U \to \mathbb{R}$ such that the restriction $\hat{f}|_{A \cap U}$ is equal to $f|_{A \cap U}$. The function f is said to be *analytic on* A if and only if it is analytic at every point in A.

Let

$$f(x_1, \ldots, x_{n-1}, t) = f_0(x_1, \ldots, x_{n-1}) + \cdots + f_d(x_1, \ldots, x_{n-1})t^d$$

be a polynomial in $(x, t) \in \mathbb{R}^{n-1} \times \mathbb{R}$ with coefficients in $\mathbb{R}$. For each $x \in \mathbb{R}^{n-1}$, we denote by f_x the polynomial $\mathbb{R} \to \mathbb{R}, t \mapsto f(x, t)$, viewed as a polynomial in the variable t, with coefficients in $\mathbb{R}[x_1, \ldots, x_{n-1}]$, the ring of polynomials in $x_1, \ldots, x_{n-1}$ with coefficients in $\mathbb{R}$.

Lemma 1.2. *Let* $A \subset \mathbb{R}^{n-1}$ *be a connected semi-algebraic set such that, for every point* $x \in A$, *the polynomial* $f_x \colon \mathbb{R}^n \to \mathbb{R}, t \mapsto f_x(t)$, *has degree* d *and exactly* k *distinct complex roots. Then there are* $l \leq k$ *analytic functions* $\xi_1 < \cdots < \xi_l \colon A \to \mathbb{R}$ *such that for every* $x \in A$, *the set of real roots of* $f_x(t)$ *is exactly* $\{\xi_1(x), \ldots, \xi_l(x)\}$.

Proof. Let $a \in A$ and $t_1(a), \ldots, t_k(a)$ be the distinct complex roots of $f_a(t)$. Let $B_i = \{t \in \mathbb{C} \mid |t - t_i(a)| < \epsilon\}$ for so small $\epsilon > 0$ such that $B_i \cap B_j = \emptyset$ for $i \neq j$ and $B_i \cap \mathbb{R} = \emptyset$ if $t_i(a) \notin \mathbb{R}$. Choose $\delta > 0$ sufficiently small such that if $x \in U(\delta) := \{x \in \mathbb{R}^{n-1} \mid \|x - a\| < \delta\}$, then $|f(x, t) - f(a, t)| < |f(a, t)|$ on the circle $S_i := \partial B_i$. Let

$$f_\theta(x, t) := f(a, t) - \theta(f(x, t) - f(a, t)), \quad \text{for } \theta \in [0, 1].$$

Then for all $x \in U(\delta)$ and all $t \in S_i$,

$$f_\theta(x, t) \geq |f(a, t)| - \theta |f(x, t) - f(a, t)|$$
$$\geq |f(a, t)| - |f(x, t) - f(a, t)| > 0.$$

Hence, for $\theta \in [0, 1]$ and $x \in U(\delta)$, the total number $m_{\theta,i}$ of roots of $f_\theta(x, t)$ inside B_i is given by

$$m_{\theta,i} = \frac{1}{2\pi\sqrt{-1}} \int_{S_i} \frac{\frac{d}{dt} f_\theta(x, t)}{f_\theta(x, t)} dt.$$

But the integral on the right-hand side is continuous in θ and is integer-valued, therefore $m_{\theta,i} = m_i$ for all $\theta \in [0, 1]$, where m_i is the multiplicity of the root $t_i(a)$.

Let $a' \in A \cap U(\delta)$. Then in each B_i, there exists at least one root $t_i(a')$ of $f_{a'}(t)$. Since $a' \in A$, the number of distinct roots of $f_{a'}(t)$ is also equal to k. Hence $t_i(a')$ is the unique root of $f_{a'}(t)$ in B_i and the multiplicity of the root $t(a')$ of $f_{a'}(t)$ is also equal to m_i. Since

$$d = m_1 + \cdots + m_k,$$

the set of roots of $f_{a'}(t)$ is exactly $\{t_1(a'), \ldots, t_k(a')\}$. If $t_i(a) \notin \mathbb{R}$, then $t_i(a') \notin \mathbb{R}$, since $B_i \cap \mathbb{R} = \emptyset$. Assume that $t_i(a)$ is a real root of $f_a(t)$, then $t_i(a')$ is also a real root of $f_{a'}(t)$, otherwise B_i contains both $t_i(a')$ and its image by conjugation, which is impossible.

We have shown that on the set $A \cap U(\delta)$, the number of distinct real roots of $f_x(t)$ is constant. Let l be this number and $\xi_1(x) < \cdots < \xi_l(x)$ be the real roots of $f_x(t)$, $x \in A \cap U(\delta)$. For each $i = 1, \ldots, l$, the multiplicity of $\xi_i(x)$, as a root of $f_x(t)$, $x \in A \cap U(\delta)$, is equal to m_i. Hence, each $\xi_i(x)$, $x \in A \cap U(\delta)$, is a simple real root of the $(m_i - 1)$th derivative of $f(x, t)$ with respect to the variable t. Then, it follows from the Implicit Function Theorem that $\xi_i \colon A \cap U(\delta) \to \mathbb{R}$ is an analytic function.

Since $\xi_i \colon A \cap U(\delta) \to \mathbb{R}$ is analytic, hence continuous, it follows from the connectedness of A that each $\xi_i(x)$ has constant multiplicity on A. Thus, we obtain l functions $\xi_1(x) < \cdots < \xi_l(x) \colon A \to \mathbb{R}$, each of them is analytic at any point $x \in A$ and the lemma is proved. $\square$

Lemma 1.3 (Thom's lemma). *Let $f_1, \ldots, f_s \in \mathbb{R}[t]$ be a finite family of non-zero polynomials, which is closed under derivation (i.e. if the derivative f_i' is non-zero, then there is j such that $f_i' = f_j$). For $(*_1, \ldots, *_s) \in \{<, =, >\}^s$, let $V_* \subset \mathbb{R}$ be defined by*

$$V_* := \{t \in \mathbb{R} \mid f_j(x) *_j 0, j = 1, \ldots, s\}.$$

Then V_* is a connected semi-algebraic subset in $\mathbb{R}$, i.e. either V_* is empty, or V_* is a point, or V_* is an open interval.

Proof. We will prove by induction on s. For $s = 1$, the only polynomial must be non-zero constant, hence $V_* = \emptyset$ or $V = \mathbb{R}$.

Assume that $s > 1$. Without loss of generality, we may assume that $\deg f_s = \max\{\deg f_i, i = 1, \ldots, s\}$. Put

$$V_{*'} = \{t \in \mathbb{R} \mid f_j(t) *_j 0, j = 1, \ldots, s - 1\}.$$

Since the family $f_1, \ldots, f_{s-1}$ is still closed under derivation, it follows from the inductive hypothesis that $V_{*'}$ is either empty or a point, or an open interval. If $V_{*'}$ is empty or a point, so is V_*, since $V_* = V_{*'} \cap \{t \in \mathbb{R} \mid f_s *_k 0\}$. Assume that $V_{*'}$ is an open interval, then f_s is either strictly monotone or constant on $V_{*'}$, since the sign of f_s' is constant on $V_{*'}$. In each case, one can easily verify that V_* is either empty, or a point, or an open interval. $\square$

Definition 1.4. Let $A \subset \mathbb{R}^n$ be a semi-algebraic set. A map $f \colon A \to \mathbb{R}^m$ is called *semi-algebraic* if its graph is a semi-algebraic subset of $\mathbb{R}^n \times \mathbb{R}^m$.

Lemma 1.4. *Let $A \subset \mathbb{R}^n$ be a semi-algebraic set and $f \colon A \to \mathbb{R}$ a semi-algebraic function. Then there is a non-zero polynomial $P \colon \mathbb{R}^n \times \mathbb{R} \to \mathbb{R}$ such that $P(x, f(x)) = 0$ for all $x \in A$.*

Proof. Since f is semi-algebraic, the graph of f is of the form

$$\bigcup_{i=1}^{s} \{(x, y) \in \mathbb{R}^n \times \mathbb{R} \mid g_i(x, y) = 0, \ h_{i_j}(x, y) > 0, \ j = 1, \ldots, k_i\}$$

for some polynomials g_i and h_{i_j}.

Since the graph of f cannot contain a non-empty open subset of $\mathbb{R}^{n+1}$, at least one among the g_i is non-zero. We can then take P to be the product of these non-zero polynomials. $\square$

1.3   Cell decomposition

1.3.1   *Definition of analytic cell decompositions*

Analytic cells are non-empty semi-algebraic sets of an especially simple nature. They are defined inductively as follows:

- Analytic cells in $\mathbb{R}$ are points $\{c\}$ or open intervals (a, b), $-\infty \leq a < b \leq +\infty$.
- Let $C \subset \mathbb{R}^{n-1}$ be an analytic cell. If $f, g \colon C \to \mathbb{R}$ are analytic semi-algebraic functions such that $f < g$ on C, then the *cylinder*

$$(f, g) := \{(x, t) \in C \times \mathbb{R} \mid f(x) < t < g(x)\}$$

as well as

$$(-\infty, f) := \{(x, t) \in C \times \mathbb{R} \mid -\infty < t < f(x)\}$$

and

$$(g, +\infty) := \{(x, t) \in C \times \mathbb{R} \mid g(x) < t < +\infty\}$$

are analytic cells in $\mathbb{R}^n$. If $f \colon C \to \mathbb{R}$ is an analytic semi-algebraic function, then its *graph*

$$\operatorname{graph} f := \{(x, t) \in C \times \mathbb{R} \mid t = f(x)\}$$

is an analytic cell in $\mathbb{R}^n$. Finally, $C \times \mathbb{R} \subset \mathbb{R}^n$ is an analytic cell.
- By convention, we also consider the one-point set $\mathbb{R}^0$ as an analytic cell.

An *analytic cell decomposition* of $\mathbb{R}^n$ is defined by induction on n:

- An analytic cell decomposition of $\mathbb{R}$ is a finite collection of open intervals and points:

$$\{(-\infty, a_1), \{a_1\}, (a_1, a_2), \ldots, \{a_k\}, (a_k, +\infty)\},$$

where $a_1 < a_2 < \cdots < a_k$ are points in $\mathbb{R}$.
- Assuming that the class of analytic cell decompositions of $\mathbb{R}^{n-1}$ has been defined, an analytic cell decomposition of $\mathbb{R}^n$ is a finite partition $\mathcal{P}$ of $\mathbb{R}^n$ into analytic cells such that

$$\pi(\mathcal{P}) := \{\pi(C) \mid C \in \mathcal{P}\}$$

is an analytic cell decomposition of $\mathbb{R}^{n-1}$, where $\pi\colon \mathbb{R}^n \to \mathbb{R}^{n-1}$ is the projection on the first $(n-1)$ coordinates.

We say that a decomposition $\mathcal{P}$ of $\mathbb{R}^n$ *partitions* a set $S \subset \mathbb{R}^n$ if S is a disjoint union of cells in $\mathcal{P}$.

The following two results give important topological and analytical properties of semi-algebraic sets.

Theorem 1.1 (Analytic Cell Decomposition). (I_n) *Let $S_1, \ldots,$ S_k be semi-algebraic subsets of $\mathbb{R}^n$. Then there is an analytic cell decomposition of $\mathbb{R}^n$ partitioning each S_i.*

Theorem 1.2 (Finiteness of the number of connected components). (II_n) *Every semi-algebraic set has a finite number of connected components and each such component is semi-algebraic.*

1.3.2 *Proof of Theorems 1.1 and 1.2*

We will prove Theorems 1.1 and 1.2 simultaneously by induction on n. We will show that

$$(I_{n-1}) + (II_{n-1}) \Rightarrow (I_n)$$

and

$$(I_n) \Rightarrow (II_n).$$

For $n = 1$, the statements (I_1) and (II_1) hold, since a semi-algebraic subset of $\mathbb{R}$ is a finite union of points and open intervals.

Assume that (I_{n-1}) and (II_{n-1}) hold; we prove that (I_n) also holds.

Step 1: *Construction of cell decomposition of $\mathbb{R}^n$, partitions $S_1, \ldots, S_k$.*

We write the coordinate of $\mathbb{R}^n$ as $(x, t) \in \mathbb{R}^{n-1} \times \mathbb{R}$. Let S be one of $S_1, \ldots, S_k$. Then S can be represented by

$$S = \bigcup_{i=1}^{p_S} \{(x, t) \in \mathbb{R}^n \mid f_i^S(x, t) = 0, \ g_{ij}^S(x, t) > 0, \ j = 1, \ldots, l_i^S\}$$

for some polynomials f_i^S, g_{ij}^S. Put

$$\mathcal{A}_S := \{f_i^S, g_{ij}, \ i = 1, \ldots, p_S, j = 1, \ldots, l_i^S\},$$

and

$$\mathcal{A} := \bigcup_{i=1}^{k} \mathcal{A}_{S_i}.$$

For $h \in \mathcal{A}$, let $d_h := \max\{\ell \mid \frac{\partial^\ell h}{\partial t^\ell} \not\equiv 0\}$. Put

$$\mathcal{B} := \mathcal{A} \cup \bigcup_{i=1}^{k} \bigcup_{h \in \mathcal{A}_{S_i}} \left\{ \frac{\partial^\ell h}{\partial t^\ell}, \ 1 \leq \ell \leq d_h \right\}.$$

Then $\mathcal{B}$ is a finite family. We denote the elements of $\mathcal{B}$ by $g_1, \ldots, g_q$, i.e.

$$\mathcal{B} = \{g_1, \ldots, g_q\} \subset \mathbb{R}[x, t].$$

For each subset ψ of $\{1, \ldots, q\}$, put

$$g_\psi := \prod_{j \in \psi} g_j.$$

We denote by $g_{\psi,x}(t)$ the polynomial $g_\psi(x, t)$, viewed as a polynomial in the variable t. Let $d_\psi := \deg g_{\psi,x}$. For each $l \leq d_\psi$ and $1 \leq k \leq l$, put

$$\mathcal{P}_{\psi,k}^l := \{x \in \mathbb{R}^{n-1} \mid \deg g_{\psi,x} = l \ \text{and}$$
$$g_{\psi,x}(t) \ \text{has exactly } k \text{ distinct complex roots}\}.$$

It follows from Lemma 1.1 that $\mathcal{P}_{\psi,k}^l$ is a semi-algebraic subset of $\mathbb{R}^{n-1}$.

Let $\mathcal{P}_\psi$ denote the family of all sets $\mathcal{P}_{\psi,k}^l, 1 \leq k \leq l \leq d_\psi$. Then it is not hard to see that $\mathcal{P}_\psi$ is a finite partition of $\mathbb{R}^{n-1}$.

Put $\mathcal{P} := \bigcap_{\psi \subset \{1,\ldots,q\}} \mathcal{P}_\psi$ — the intersection of all partitions $\mathcal{P}_\psi$. By definition, each element V of $\mathcal{P}$ is of the form

$$V = \bigcap V_\psi,$$

where $V_\psi \in \mathcal{P}_\psi$ and ψ runs through all the subsets of $\{1, \ldots, q\}$. Clearly, $\mathcal{P}$ is again a finite partition of $\mathbb{R}^{n-1}$ and each element of $\mathcal{P}$ is a semi-algebraic subset of $\mathbb{R}^n$. Let us denote elements of $\mathcal{P}$ by

$V_1, \ldots, V_r$, i.e.

$$P = \{V_1, \ldots, V_r\}.$$

Finally, we denote by $\mathcal{D}^{n-1}$ the family of all connected components of V_i, $i = 1, \ldots, r$. By the inductive hypothesis, (II_{n-1}) holds, $\mathcal{D}^{n-1}$ is a finite partition of $\mathbb{R}^{n-1}$,

$$\mathcal{D}^{n-1} = \{W_1, \ldots, W_u\},$$

where each W_i is a connected semi-algebraic subset of $\mathbb{R}^{n-1}$.

Moreover, since (I_{n-1}) holds by induction, we can assume that each W_i, $i = 1, \ldots, u$, is an analytic cell in $\mathbb{R}^{n-1}$, i.e. $\mathcal{D}^{n-1}$ is an analytic cell decomposition of $\mathbb{R}^{n-1}$.

For each $W \in \mathcal{D}^{n-1}$, let $G_W(x, t)$ be the product of $g_j(x, t) \in \mathcal{B}$ which are non-zero polynomials (in the variable t) for all $x \in W$. It follows from the construction of $\mathcal{D}^{n-1}$ that the degree and the number of distinct complex roots of $G_W(x, t)$ are constant for all $x \in W$. Then, by Lemma 1.2, there are analytic functions

$$\xi_1^W < \cdots < \xi_{l_W}^W : W \to \mathbb{R}$$

such that for every $x \in W$, the set of real roots of $G_W(x, t)$ is exactly $\{\xi_1^W(x), \ldots, \xi_{l_W}^W(x)\}$. Clearly, we have

$$W \times \mathbb{R} = \left(-\infty, \xi_1^W\right) \cup \Gamma\left(\xi_1^W\right) \cup \cdots \cup \Gamma\left(\xi_{l_W}^W\right) \cup \left(\xi_{l_W}^W, +\infty\right),$$

where $\Gamma(\xi_i^W)$ stands for the graph of ξ_i^W. Hence, the family of cylinders and graphs

$$\mathcal{D}^n := \bigcup_{W \in \mathcal{D}^{n-1}} \left\{ \left(-\infty, \xi_i^W\right), \Gamma\left(\xi_i^W\right), \left(\xi_i^W, \xi_{i+1}^W\right), \right.$$

$$i = 1, \ldots, l_W - 1, \left. \left(\xi_{l_W}^W, +\infty\right) \right\}$$

forms a finite partition of $\mathbb{R}^n$ and

$$\mathcal{D}^{n-1} = \{\pi(\widetilde{W}) \mid \widetilde{W} \in \mathcal{D}^n\},$$

where $\pi \colon \mathbb{R}^n \to \mathbb{R}^{n-1}$, $(x, t) \mapsto x$, is the natural projection. Since each $W = \pi(\widetilde{W})$ is connected, it is easy to see that each $\widetilde{W} \in \mathcal{D}^n$ is also connected.

Step 2: Proving that each $\widetilde{W} \in \mathcal{D}^n$ is a semi-algebraic subset of $\mathbb{R}^n$.

Let $\widetilde{W} \in \mathcal{D}^n$ and $W = \pi(\widetilde{W})$. As before, let us denote $G_W(x,t)$ the product of elements of $\mathcal{B} = \{g_1(x,t), \ldots, g_q(x,t)\}$ which are non-zero polynomials (in the variable t) for all $x \in W$. Let $g \in \mathcal{B}$. If g is not a factor of $G_W(x,t)$, then $g(x,t) = 0$ for all $(x,t) \in \widetilde{W}$. If $g(x,t)$ is a factor of $G_W(x,t)$, then the real roots of $g(x,t)$ over W form a subset of the set $\{\xi_1^W(x), \ldots, \xi_{l_W}^W(x)\}$. Hence $g(x,t)$ has constant sign on $\widetilde{W}$.

We define the map $\tau \colon \{1, \ldots, q\} \to \{>, <, =\}$ by

$$\tau(j) := \begin{cases} > & \text{if } g_j(x,t) > 0 \quad \text{for } (x,t) \in \widetilde{W}, \\ < & \text{if } g_j(x,t) < 0 \quad \text{for } (x,t) \in \widetilde{W}, \\ = & \text{if } g_j(x,t) = 0 \quad \text{for } (x,t) \in \widetilde{W}. \end{cases}$$

Put

$$\widetilde{W}' = \{(x,t) \in W \times \mathbb{R} \mid g_j(x,t) \, \tau(j) \, 0, \ j = 1, \ldots, q\}.$$

Then $\widetilde{W}'$ is a semi-algebraic set. We will show that $\widetilde{W}' = \widetilde{W}$. Clearly, $\widetilde{W} \subset \widetilde{W}'$. Assume, by contradiction, there was a point $(a,t) \in \widetilde{W}' \backslash \widetilde{W}$, then for some t' there would be $(a,t') \in \widetilde{W}$. By Thom's lemma (Lemma 1.3), the set $(\{a\} \times \mathbb{R}) \cap \widetilde{W}'$ is connected and semi-algebraic, hence $\{a\} \times [t,t'] \subset \widetilde{W}'$. Whatever $\widetilde{W}$ is of the form (ξ_i^W, ξ_{i+1}^W), $(-\infty, \xi_1^W)$, $(\xi_{l_W}^W, +\infty)$ or $\Gamma(\xi_i^W)$, $i = 1, \ldots, l_W - 1$, there exist $t_1, t_2 \in [t,t']$ such that $G_W(a,t_1) = 0$ while $G_W(a,t_2) \neq 0$. But the sign of each $g \in \mathcal{B}$ is constant on $\widetilde{W}$, it is impossible.

Now, since each $\widetilde{W} \in \mathcal{D}^n$ is semi-algebraic, $\xi_1^W(x), \ldots, \xi_{l_W}^W$ are semi-algebraic functions. Thus, $\mathcal{D}^n$ is an analytic cell decomposition of $\mathbb{R}^n$. By construction, we can see that $\mathcal{D}^n$ partitions the sets $S_1, \ldots, S_k$. The statement (I_n) is proved.

Proof of $(\text{I}_n) \Rightarrow (\text{II}_n)$: Let $S \subset \mathbb{R}^n$ be a semi-algebraic set. We take, by (I_n), a cell decomposition of $\mathbb{R}^n$ which partitions S. Then S and each of its connected components are disjoint unions of a finite number of cells of the decomposition. Since each cell is a connected semi-algebraic set, the statement (II_n) holds.

Theorems 1.1 and 1.2 are proved.

1.3.3 *Properties of cells*

Definition 1.5. A non-empty subset A of $\mathbb{R}^n$ is an *analytic k-dimensional submanifold* of $\mathbb{R}^n$, where $1 \leq k \leq n$, if for each point p in A there exists an open neighborhood W of p in $\mathbb{R}^n$ and an open set U in $\mathbb{R}^k$ such that after some permutation of coordinates, we have

$$A \cap W = \operatorname{graph} \phi := \{(u, \phi(u)) \mid u \in U\},$$

where $\phi \colon U \to \mathbb{R}^{n-k}$ is an analytic map.

By convention, a point in $\mathbb{R}^n$ is an analytic 0-dimensional submanifold of $\mathbb{R}^n$.

Proposition 1.2. *Let $\mathcal{D}$ be an analytic cell decomposition of $\mathbb{R}^n$ and $C \in \mathcal{D}$ be a cell. Then C is an analytic submanifold of $\mathbb{R}^n$.*

Proof. By induction on n. For $n = 1$, the statement of the proposition is trivial, since cells in $\mathbb{R}$ are either points or open intervals.

Assume that $n > 1$ and let $C' = \pi(C)$, where $\pi \colon \mathbb{R}^n \to \mathbb{R}^{n-1}$ is the projection on the first $(n-1)$ coordinates. Then C' is an analytic cell of the decomposition

$$\pi(\mathcal{D}) := \{\pi(W) \mid W \in \mathcal{D}\}.$$

By the inductive hypothesis, C' is an analytic submanifold of $\mathbb{R}^{n-1}$. Let $k := \dim C'$. One can easily show that

(a) If C is the graph of an analytic function $\xi \colon C' \to \mathbb{R}$, then C is an analytic k-dimensional submanifold of $\mathbb{R}^n$;
(b) If C is the cylinder $C = (\xi_1, \xi_2)$, or $C = (-\infty, \xi_1)$, or $C = (\xi_2, +\infty)$, where $\xi_1, \xi_2 \colon C' \to \mathbb{R}$ are analytic functions satisfying $\xi_1 < \xi_2$ on C', then C is an analytic $(k + 1)$-dimensional submanifold of $\mathbb{R}^n$.

The proposition is proved. $\qquad\square$

Let us denote by $(0,1)^d \subset \mathbb{R}^d$ the open hypercube in $\mathbb{R}^d$ (with $(0,1)^0$ being a point).

Proposition 1.3. *Let $\mathcal{D}$ be an analytic decomposition of $\mathbb{R}^n$ and C be a cell of $\mathcal{D}$. Then there exist $d \in \{0, \ldots, n\}$ and a semi-algebraic map*

$$f \colon C \to (0, 1)^d \subset \mathbb{R}^d$$

such that f is an analytic isomorphism (i.e. both f and f^{-1} are analytic maps).

Proof. By induction on n. If $n = 1$, a cell in $\mathbb{R}$ is either a point or an open interval, and the proposition holds.

Assume that the proposition is true for $n - 1$. Let $C' = \pi(C)$, where π is the projection on the first $(n-1)$ coordinates of $\mathbb{R}^n$. Then C' is an analytic cell in $\mathbb{R}^{n-1}$ and C is either a graph or a cylinder over C'.

If C is the graph of an analytic semi-algebraic function $\xi \colon C' \to \mathbb{R}$, then the map

$$C' \to C, \quad x' \mapsto (x', \xi(x')),$$

is semi-algebraic and gives an analytic isomorphism between C and C'.

Assume now that C is a cylinder (ξ_1, ξ_2). We define a map

$$f \colon C' \times (0, 1) \to C, \quad (x', t) \mapsto f(x', t),$$

as follows:

$$f(x', t) := \begin{cases} (x', (1 - t)\xi_1(x') + t\xi_2(x')) & \text{if } \xi_1 \neq -\infty \text{ and } \xi_2 \neq +\infty, \\[2mm] \left(x', \dfrac{t - 1}{t} + \xi_2(x')\right) & \text{if } \xi_1 = -\infty \text{ and } \xi_2 \neq +\infty, \\[2mm] \left(x', \dfrac{t}{1 - t} + \xi_1(x')\right) & \text{if } \xi_1 \neq -\infty \text{ and } \xi_2 = +\infty, \\[2mm] \left(x', \dfrac{-1}{t} + \dfrac{1}{1 - t}\right) & \text{if } \xi_1 = -\infty \text{ and } \xi_2 = +\infty. \end{cases}$$

It is easy to see that the map f is semi-algebraic and gives us an analytic isomorphism between $C' \times (0, 1)$ and C.

The proposition follows from the inductive hypothesis. $\square$

1.4   Stratification of semi-algebraic sets

A *semi-algebraic homeomorphism* $f\colon A \to B$ is a bijective continuous semi-algebraic map from A onto B such that the inverse $f^{-1}\colon B \to A$ is continuous. The following strong version of cell decomposition of semi-algebraic sets is also true.

Theorem 1.3. *Let $S \subset \mathbb{R}^n$ be a semi-algebraic set and $S_1, \ldots, S_k$ be finitely many semi-algebraic subsets of S. Then S can be decomposed as a finite disjoint union $S = \bigcup_{i=1}^{p} C_i$, where*

- *every C_i is semi-algebraically homeomorphic to an open hypercube $(0,1)^{d_i}$;*
- *the closure $\overline{C}_i$ of C_i in S is the union of C_i and some C_j, with $j \neq i$ and $d_j < d_i$;*
- *every S_k is the union of some C_i.*

A decomposition $S = \bigcup_{i=1}^{p} C_i$ as in the above theorem is called a *stratification* of S and the C_i are called *strata* of this stratification.

1.5   Dimension of semi-algebraic sets via cell decomposition

Let C be a cell in $\mathbb{R}^n$. By Theorem 1.3, C is semi-algebraically homeomorphic to an open hypercube $(0,1)^d \subseteq \mathbb{R}^n$ for some d, and then we put $\dim C := d$.

Let $S \subset \mathbb{R}^n$ be a semi-algebraic set. By Analytic Cell Decomposition Theorem (Theorem 1.1), we can write $S = \bigcup_{i=1}^{p} C_i$ as a disjoint union of cells C_i. The *dimension* of S is, by definition,

$$\dim S := \max_{i=1,\ldots,p} \dim C_i.$$

One can show that if $S = \bigcup_{j=1}^{q} D_j$ is another decomposition of S, then

$$\max_{i=1,\ldots,p} \dim C_i = \max_{j=1,\ldots,q} \dim D_j.$$

Hence, the dimension of S is well defined and independent of the decomposition of S. As it follows from the definition, $\dim S = n$ if and only if S contains a non-empty open set.

Proposition 1.4. *Let $S \subset \mathbb{R}^n$ be a semi-algebraic set. Then*

(i) $\dim \overline{S} = \dim S$ *and* $\dim(\overline{S} \backslash S) < \dim S$.

(ii) *Let $f \colon S \to \mathbb{R}^m$ be a semi-algebraic map. Then $\dim f(S) \leq \dim S$.*

Proof. (i) Let $S = \bigcup_{i=1}^{p} C_i$ be a stratification of S (see Theorem 1.3). Then for each strata $C_i, i = 1, \ldots, p$, $\dim \overline{C}_i = \dim C_i$ and $\dim(\overline{C}_i \backslash C_i) < \dim C_i$. Hence, we have (i).

(ii) We first prove the claim: *Let $A \subset \mathbb{R}^n \times \mathbb{R}^m$ be a semi-algebraic set and $\pi \colon \mathbb{R}^n \times \mathbb{R}^m \to \mathbb{R}^n$ be the natural projection. Then $\dim \pi(A) \leq \dim A$.*

In fact, if $m = 1$, then A is a union of graphs and cylinders, hence the claim is trivial. The case $m > 1$ is proved easily by induction on m.

By the above claim, then $\dim f(S) \leq \dim(\text{graph } f) = \dim S$. $\square$

Recall that a subset $S \subset \mathbb{R}^n$ has *measure zero* if for every $\epsilon > 0$, there exists a countable family of rectangular solids covering S, the sum of whole measures is less than ϵ. Let us note that, for a subset S of $\mathbb{R}^n$, the following notions are very different: S is full measure, meaning its complement has measure zero, and S is topologically generic, meaning it contains a countable intersection of dense and open sets. However for semi-algebraic sets, the situation simplifies drastically as shown in the following proposition.

Proposition 1.5. *Let $S \subset \mathbb{R}^n$ be a semi-algebraic set. Then the following properties are equivalent:*

(i) S *is dense in* $\mathbb{R}^n$.

(ii) $\mathbb{R}^n \backslash S$ *has measure zero.*

(iii) $\dim(\mathbb{R}^n \backslash S) < n$.

Proof. (i) $\Rightarrow$ (iii) and (ii) $\Rightarrow$ (iii): By contradiction, suppose that $\dim(\mathbb{R}^n \backslash S) = n$. The set $\mathbb{R}^n \backslash S$ is semi-algebraic (because S is semi-algebraic) and hence it contains a non-empty open set. This implies that $\mathbb{R}^n \backslash S$ has no measure zero and $\overline{S} \neq \mathbb{R}^n$, which contradict the assumptions.

(iii) $\Rightarrow$ (i) and (ii): Let $S \subset \mathbb{R}^n$ be a semi-algebraic set such that $\dim(\mathbb{R}^n \backslash S) < n$.

Take any $x^* \notin S$. By assumption, $U \cap S \neq \emptyset$, for all open sets U containing x^*. This implies that $x^* \in \overline{S}$ and hence S is dense in $\mathbb{R}^n$.

On the other hand, by Analytic Cell Decomposition Theorem (Theorem 1.1), we can write $\mathbb{R}^n \backslash S$ as a disjoint union of finitely many cells C_i with $\dim C_i < n$. Each C_i has measure zero and so does the finite union of C_i. Therefore, $\mathbb{R}^n \backslash S$ has measure zero. $\square$

Theorem 1.4 (Uniform bound). *For any semi-algebraic set $S \subset \mathbb{R}^n \times \mathbb{R}^m$, there exists an integer $N \geq 1$ such that for all $x \in \mathbb{R}^n$, the number of connected components of the set $S_x := \{y \in \mathbb{R}^m \mid (x, y) \in S\}$ is bounded from above by N.*

Proof. Let $\pi \colon \mathbb{R}^n \times \mathbb{R}^m \to \mathbb{R}^n, (x, y) \mapsto x$, be the projection on the first n coordinates. By Theorem 1.1, there is a cell decomposition $\mathcal{D}$ of $\mathbb{R}^n \times \mathbb{R}^m$ which partitions S such that

$$\pi(\mathcal{D}) := \{\pi(C) \mid C \in \mathcal{D}\}$$

is a cell decomposition of $\mathbb{R}^n$. Then, we can write

$$\pi(S) = C_1' \cup \cdots \cup C_p',$$

for some cells $C_i' \in \pi(\mathcal{D}), i = 1, \dots, p$.

For each $x \in C_i', i = 1, \dots, p$, the number of connected components of the set

$$S_x := \{t \in \mathbb{R} \mid (x, t) \in S\}$$

is bounded from above by the number $\#\{C \in \mathcal{D} \mid \pi(C) = C_i'\}$. $\square$

1.6 The Tarski–Seidenberg theorem

Theorem 1.5 (Tarski–Seidenberg theorem). *Let S be a semi-algebraic subset of $\mathbb{R}^n \times \mathbb{R}^m$ and $\pi \colon \mathbb{R}^n \times \mathbb{R}^m \to \mathbb{R}^n, (x, y) \mapsto x$, be the projection map. Then $\pi(S)$ is a semi-algebraic subset of $\mathbb{R}^n$.*

Proof. By induction on m. Let $m = 1$. By Theorem 1.1, there is a finite partition $\mathcal{D}$ of $\mathbb{R}^n$ such that if $C \in \mathcal{D}$ then C is a semi-algebraic subset of $\mathbb{R}^n$ and

$$\pi(S) = \bigcup \{ C \in \mathcal{D} \mid (C \times \mathbb{R}) \cap S \neq \emptyset \}.$$

Hence, $\pi(S)$ is semi-algebraic.

Assume now that the theorem holds for $m-1$. Identifying $\mathbb{R}^n \times \mathbb{R}^m$ with $\mathbb{R}^{n+1} \times \mathbb{R}^{m-1}$, we consider the projections $\pi_1 \colon \mathbb{R}^{n+1} \times \mathbb{R}^{m-1} \to \mathbb{R}^{n+1}$ and $\pi_2 \colon \mathbb{R}^{n+1} \to \mathbb{R}^n$. Clearly, $\pi = \pi_2 \circ \pi_1$. By the inductive hypothesis, $\pi_1(S)$ is semi-algebraic and then $\pi(S) = \pi_2(\pi_1(S))$ is semi-algebraic. $\qquad\square$

1.6.1 *Logical formulation of the Tarski–Seidenberg theorem*

We will use logical formulas in a familiar way, as convenient descriptions or definitions of sets and functions.

To illustrate this notational use of formulas, let x, y, z be variables ranging over sets X, Y, and Z, respectively and let $\phi(x, y, z)$ and $\psi(x, y, z)$ be *formulas* (conditions on (x, y, z)) defining sets

$$\Phi := \{ (x, y, z) \in X \times Y \times Z \mid \phi(x, y, z) \text{ holds} \},$$

$$\Psi := \{ (x, y, z) \in X \times Y \times Z \mid \psi(x, y, z) \text{ holds} \}.$$

Then we can construct new formulas:

- the *disjunction* of ϕ and ψ, denoted by $\phi \vee \psi$, defines the set $\Phi \cup \Psi$;
- the *conjunction* of ϕ and ψ, denoted by $\phi \wedge \psi$, defines $\Phi \cap \Psi$;
- the *negation* of ϕ, denoted by $\neg\phi$, defines the complement $X \times Y \times Z \backslash \Phi$;
- the *existential quantification* over z of $\phi(x, y, z)$, denoted by $\exists z \phi(x, y, z)$, defines the set $\{ (x, y) \in X \times Y \mid \text{there exists } z \in Z$ such that $\phi(x, y, z)$ holds$\}$.
  Note that if $\pi \colon X \times Y \times Z \to X \times Y$ is the projection map, then the formula $\exists z \phi(x, y, z)$ defines the set $\pi(\Phi)$;
- the *universal quantification* over z of $\phi(x, y, z)$, $\forall z \phi(x, y, z)$, defines the set $\{ (x, y) \in X \times Y \mid \text{for all } z \in Z$ the condition $\phi(x, y, z)$ holds$\}$.

Clearly, we have

- $\forall z \phi(x, y, z)$ and $\neg \exists z \neg \phi(x, y, z)$ defines the same set;
- the implication $\phi(x, y, z) \to \psi(x, y, z)$ can be considered as the formula $(\neg \phi(x, y, z) \vee \psi(x, y, z))$.

Definition 1.6. A *first-order formula* in the language of real closed fields is obtained as follows recursively:

(1) If $f \in \mathbb{R}[x_1, \ldots, x_n], n \geq 1$, then
$$f = 0 \quad \text{and} \quad f > 0$$
are first-order formulas (with free variables $x = (x_1, \ldots, x_n)$) and $\{x \in \mathbb{R}^n \mid f(x) = 0\}$ and $\{x \in \mathbb{R}^n \mid f(x) > 0\}$ are respectively the subsets of $\mathbb{R}^n$ such that the formulas $f = 0$ and $f > 0$ hold.
(2) If ϕ and ψ are first-order formulas, then $\phi \wedge \psi$ (conjunction), $\phi \vee \psi$ (disjunction) and $\neg \phi$ (negation) are also first-order formulas.
(3) If ϕ is a first-order formula and x is a variable ranging over $\mathbb{R}$, then $\exists x \phi$ and $\forall x \phi$ are first-order formulas.

The formulas obtained by using only rules (1) and (2) are called *quantifier-free formulas*.

It follows from definitions that a subset $S \subset \mathbb{R}^n$ is semi-algebraic if and only if there exists a quantifier-free formula $\phi(x_1, \ldots, x_n)$ such that
$$S = \{x = (x_1, \ldots, x_n) \in \mathbb{R}^n \mid \phi(x_1, \ldots, x_n) \text{ holds}\}.$$
With the Tarski–Seidenberg theorem, we can say more, namely, we have

Theorem 1.6 (Logical formulation of the Tarski–Seidenberg theorem). *If $\phi(x)$ is a first-order formula, then the set $\{x \in \mathbb{R}^n \mid \phi(x) \text{ holds}\}$ is semi-algebraic.*

Proof. Since $\forall x \phi$ is equivalent to the formula $\neg \exists x \neg \phi$, we need only to verify that if $\phi(x_1, \ldots, x_{n+1})$ is a quantifier-free formula, then the set
$$\{(x_1, \ldots, x_n) \in \mathbb{R}^n \mid \exists x_{n+1} \text{ such that } \phi(x_1, \ldots, x_{n+1}) \text{ holds}\}$$
is semi-algebraic. But this is a direct consequence of Theorem 1.5.

$\square$

Proposition 1.6. *The following two statements hold.*

(i) *If A and B are semi-algebraic, then $A \times B$ is also semi-algebraic.*

(ii) *The closure, the interior and the boundary of a semi-algebraic set are semi-algebraic.*

Proof. (i) Let $A := \{x \in \mathbb{R}^n \,|\, \phi(x) \text{ holds}\}$ and $B := \{y \in \mathbb{R}^m \,|\, \psi(x) \text{ holds}\}$, where ϕ and ψ are first-order formulas. Then

$$A \times B = \{(x, y) \in \mathbb{R}^n \times \mathbb{R}^m \,|\, \phi(x) \wedge \psi(y) \text{ holds}\}.$$

Hence, $A \times B$ is semi-algebraic.

  (ii) Let $S \subset \mathbb{R}^n$ be given by

$$S = \{x \in \mathbb{R}^n \,|\, \phi(x) \text{ holds}\},$$

where $\phi(x)$ is a first-order formula. Then the closure of S is given by

$$\overline{S} = \{x \in \mathbb{R}^n \,|\, \forall \epsilon > 0 \; \exists y \in \mathbb{R}^n \text{ such that } (\phi(y) \wedge \|y - x\|^2 < \epsilon) \text{ holds}\}$$

with $\|x\|^2 := \sum_{i=1}^n x_i^2$. Hence, by Theorem 1.6, $\overline{S}$ is semi-algebraic. We remark that the interior and the boundary of S are given by $\mathbb{R}^n \backslash (\overline{\mathbb{R}^n \backslash S})$ and $\overline{S} \backslash (\mathbb{R}^n \backslash (\overline{\mathbb{R}^n \backslash S}))$, respectively, and hence are semi-algebraic. $\square$

Proposition 1.7. *The following statements hold.*

(i) *Images and inverse images of semi-algebraic sets under semi-algebraic maps are semi-algebraic.*

(ii) *Compositions of semi-algebraic maps are semi-algebraic.*

(iii) *The sum and product of two semi-algebraic functions are semi-algebraic.*

Proof. (i) Let $f \colon A \to B$ be a semi-algebraic map and $S \subset A$, $T \subset B$ be semi-algebraic subsets. Let $\pi_A \colon A \times B \to A$ and $\pi_B \colon A \times B \to B$ be the projections. We have

$$f(S) = \pi_B((S \times B) \cap \operatorname{graph} f)$$

and

$$f^{-1}(T) = \pi_A((A \times T) \cap \operatorname{graph} f).$$

Hence, $f(S)$ and $f^{-1}(T)$ are semi-algebraic.

(ii) Let $f \colon A \to B$ and $g \colon B \to C$ be semi-algebraic maps. Then

$$\text{graph } (g \circ f) = \pi((\text{graph } f \times C) \cap (A \times \text{graph } g)),$$

where $\pi \colon A \times B \times C \to A \times C$ defined by $\pi(a, b, c) = (a, c)$. So $g \circ f$ is semi-algebraic.

(iii) Let $f \colon A \to \mathbb{R}$ and $g \colon A \to \mathbb{R}$ be semi-algebraic. Define the maps

$$(f, g) \colon A \to \mathbb{R}^2, \quad x \mapsto (f(x), g(x)),$$

$$+ \colon \mathbb{R}^2 \to \mathbb{R}, \quad (x, y) \mapsto x + y,$$

and

$$\times \colon \mathbb{R}^2 \to \mathbb{R}, \quad (x, y) \mapsto xy.$$

Then $f + g = + \circ (f, g)$ and $fg = \times \circ (f, g)$. Hence, $f + g$ and fg are semi-algebraic functions. $\qquad \square$

Proposition 1.8. *Let $A \subset \mathbb{R}^n$ be a non-empty semi-algebraic set. Then the distance function*

$$\text{dist}(\cdot, A) \colon \mathbb{R}^n \to \mathbb{R}, \quad x \mapsto \text{dist}(x, A) := \inf\{\|x - a\| \mid a \in A\},$$

is semi-algebraic.

Proof. The graph of the function $\text{dist}(\cdot, A)$ is

$$\{(x, t) \in \mathbb{R}^{n+1} \mid t \geq 0 \text{ and } \forall y \in A, t^2 \leq \|x - y\|^2 \text{ and}$$

$$\forall \epsilon \in \mathbb{R}, \epsilon > 0 \Rightarrow \exists y \in A, t^2 + \epsilon > \|x - y\|^2\}.$$

In the above formula, A is a semi-algebraic subset, $\|x - y\|^2$ is a polynomial function. Moreover, if ϕ and ψ are first-order formulas, then $\phi \Rightarrow \psi$ is also a first-order formula. Hence, the graph of $\text{dist}(\cdot, A)$ is defined by a first-order formula, and so it is a semi-algebraic set. $\qquad \square$

1.6.2 *Some examples of semi-algebraic sets and functions*

Example 1.2. Let $f \colon A \to \mathbb{R}$ be a semi-algebraic function. Suppose that f is bounded from below. Let $g \colon A \to \mathbb{R}^m$ be a semi-algebraic map. Then the function

$$\varphi\colon g(A) \to \mathbb{R}, \quad y \mapsto \inf_{x \in g^{-1}(y)} f(x),$$

is semi-algebraic. The proof is similar to that of Proposition 1.8.

Example 1.3. Let $f\colon \mathbb{R} \to \mathbb{R}$ be a semi-algebraic function. Then

$$\{x \in \mathbb{R} \mid f \text{ is convex in an interval around } x\}$$

is semi-algebraic.

Indeed, "f is convex in the interval around x" can be written as

$$\exists (a, b)(a < x < b \text{ and } ``f \text{ is convex in the interval } (a, b)")$$

and "f is convex in the interval (a, b)" can be written as

$$\forall (t, x_1, x_2), (a < x_1 < t < x_2 < b)$$
$$\Rightarrow f(t) \leq \frac{x_2 - t}{x_2 - x_1} f(x_1) + \frac{t - x_1}{x_2 - x_1} f(x_2).$$

Example 1.4. Let $f\colon A \subset \mathbb{R}^n \to \mathbb{R}$ be a semi-algebraic function. Then

$$\{x \in A \mid f \text{ is continuous at } x\}$$

is a semi-algebraic subset.

Indeed. "f is continuous at x" can be written as

$$\forall \epsilon > 0 \; \exists \delta > 0 \; \forall y \in A \; (\|y - x\| < \delta \Rightarrow |f(y) - f(x)| < \epsilon).$$

The same conclusion also holds for differentiability.

Remark 1.2. We emphasize the fact that quantification over real numbers only is allowed. For example, the set

$$\{(x, y) \in \mathbb{R}^2 \mid \exists n \in \mathbb{N}, y = nx\}$$

is not semi-algebraic.

1.7 Piecewise analyticity of semi-algebraic functions

Theorem 1.7 (Piecewise analyticity of semi-algebraic functions). *Given a semi-algebraic function $f\colon S \to \mathbb{R}$, where S is a semi-algebraic subset of $\mathbb{R}^n$, there is a finite partition of S into analytic semi-algebraic submanifolds $C_1, C_2, \ldots, C_p$, such that each the restriction $f|_{C_i}$ is an analytic function.*

Proof. By Theorem 1.1, there is an analytic cell decomposition $\mathcal{D}$ of $\mathbb{R}^{n+1}$ partitioning the graph of f. Since $\dim \operatorname{graph} f = \dim S \leq n$, each cell in $\mathcal{D}$ is the graph of some analytic semi-algebraic function, and then the theorem follows easily. $\qquad\square$

1.8 Monotonicity theorem

Theorem 1.8 (Monotonicity Theorem). *Let $f\colon (a,b) \to \mathbb{R}$ be a semi-algebraic function. Then there are $a = a_0 < a_1 < \cdots < a_s < a_{s+1} = b$ such that, for each $i = 0, \ldots, s$, the restriction $f|_{(a_i, a_{i+1})}$ is analytic, and either constant, or strictly increasing or strictly decreasing.*

Proof. By Theorem 1.7, we may assume that f is analytic on (a, b). We show the next claim.

Claim: *There is a subinterval of (a,b) on which f is either constant or strictly monotone.*

Proof. Let $p, q \in (a,b), p < q$, and suppose that f is not constant on (p, q). Then $f([p,q])$ contains a closed interval $[c, d]$, $c < d$. We define the function

$$g\colon [c, d] \to [p, q], \quad y \mapsto \min\{x \in [p, q] \mid f(x) = y\}.$$

It is not hard to see that g is injective and semi-algebraic, and hence it is analytic on some interval $I \subseteq [c, d]$ by Theorem 1.7. Therefore, g is strictly monotone on I. Thus, f is strictly monotone on the interval $g(I) \subseteq [p, q]$, which proves the claim. $\qquad\square$

Now, we are ready to finish the proof of the theorem. Let E be the set of all points $x \in (a,b)$ such that there is no subinterval of (a,b)

around x on which f is either constant or strictly monotone. One can show that E is semi-algebraic. By claim, E cannot contain an interval. So E is a finite set, say $E = \{a_1, \ldots, a_s\}$. Then the theorem holds with these a_i's.   □

1.9 Semi-algebraic Sard's theorem

Given a differentiable map between manifolds $f \colon X \to Y$, a point $y \in Y$ is called a *regular value* for f if either $f^{-1}(y) = \emptyset$ or the derivative map $Df(x) \colon T_xX \to T_yY$ is surjective at every point $x \in f^{-1}(y)$, where T_xX and T_yY denote the tangent spaces of X at x and of Y at y, respectively. A point $y \in Y$ that is not a regular value of f is called a *critical value*. We will denote by $K_0(f)$ the *set of critical values* of f.

The classical Sard theorem says that if f is of class C^r and if

$$r > \max\{\dim X - \dim Y, 0\},$$

then the set $K_0(f)$ has measure zero. The examples of H. Whitney (1965) and Y. Yomdin (1991) show that the theorem does not hold for maps of low smoothness. We will see that if f is a semi-algebraic map, then the Sard theorem is much stronger than the classical one: we need weaker assumptions, but we get stronger conclusions.

Theorem 1.9 (Sard's theorem). *Let $f \colon X \to Y$ be a differentiable semi-algebraic map between two semi-algebraic submanifolds of $\mathbb{R}^n$ and $\mathbb{R}^m$, respectively. Then the set $K_0(f)$ of critical values of f is a semi-algebraic subset of Y, of dimension smaller than the dimension of Y. In particular, the set $K_0(f)$ has measure zero in Y.*

Proof. We leave the checking of semi-algebraicity of $K_0(f)$ to the reader and we will prove that $\dim K_0(f) < \dim Y$.

If $\dim X < \dim Y$, then $\dim K_0(f) \le \dim f(X) \le \dim X < \dim Y$. Thus, we can assume that $\dim Y \le \dim X$. It follows from Theorem 1.7 that there exists a finite decomposition $\{C_i\}$ of X into analytic cells of $\mathbb{R}^n$ such that the restriction of f on each C_i is an analytic semi-algebraic map. Let Δ_i denote the set of critical values of $f|_{C_i}$. By the classical Sard theorem, each Δ_i has measure zero in Y. Hence, $\dim \Delta_i < \dim Y$, since Δ_i is semi-algebraic.

Let $y \in K_0(f)$. Then there exist C_i and $x \in C_i \cap f^{-1}(y)$ such that the derivative map $Df(x) \colon T_x X \to T_y Y$ is not surjective. We now show that if $\dim Y \leq \dim C_i$, then $y \in \Delta_i$. Indeed, this is true if $\dim C_i = \dim X$, since in this case, C_i is an open subset of X. Assume that $\dim Y \leq \dim C_i < \dim X$. Then, the tangent space $T_x C_i$ of C_i at x is a subspace of the space $T_x X$ and the derivative map $D(f|_{C_i})(x) \colon T_x C_i \to T_y Y$ is the restriction of the map $Df(x) \colon T_x X \to T_y Y$ to the subspace $T_x C_i$. We have

$$D(f|_{C_i})(x)(T_x C_i) \subset Df(x)(T_x X).$$

Hence, $D(f|_{C_i})(x)$ is not surjective, since $Df(x)(T_x X)$ is a proper subspace of $T_y Y$. Therefore, $y \in \Delta_i$.

We have showed that

$$K_0(f) \subset \bigcup \Delta_i \bigcup_{\dim C_i < \dim Y} f(C_i)$$

and so $\dim K_0(f) < \dim Y$. $\qquad \square$

Corollary 1.1. *Let $f \colon X \to \mathbb{R}$ be a differentiable semi-algebraic function. Then f has finitely many critical values.*

Theorem 1.10 (Sard's theorem with parameter). *Let $f \colon P \times X \to Y$ be a differentiable semi-algebraic map between semi-algebraic submanifolds. If $y \in Y$ is a regular value of f, then there exists a semi-algebraic set $\Sigma \subset P$ of dimension smaller than the dimension of Y such that, for every $p \in P \backslash \Sigma$, y is a regular value of the map $f_p \colon X \to Y, x \mapsto f(p, x)$.*

Proof. The theorem is a direct consequence of the following two facts:

(a) for almost every $p \in P$, y is a regular value of the map f_p, and
(b) the set of points $p \in P$ such that y is not a regular value of the map f_p is a semi-algebraic subset of P.

Indeed, without loss of generality, we can assume that $f^{-1}(y) \neq \emptyset$. From the Implicit Function Theorem, it follows that the preimage $f^{-1}(y)$ is a submanifold in $P \times X$. Let $\pi \colon P \times X \to P, (p,x) \mapsto p$, be the natural projection map. By Theorem 1.9, the set of critical values $\Sigma := K_0(\pi)$ of the restriction map $\pi \colon f^{-1}(y) \to P$ is a semi-algebraic subset in P of dimension smaller than the dimension of P.

Now, let p be an arbitrary point in $P \backslash \Sigma$. We shall prove that y is a regular value of f_p. In fact, let $x \in X$ be such that $f_p(x) = y$. Given any vector $a \in T_y Y$, we want to exhibit a vector $b \in T_x X$ such that

$$Df_p(x)(b) = a.$$

Because y is a regular value of f, we know that

$$Df(p, x)\left(T_{(p,x)}(P \times X)\right) = T_y Y.$$

Hence, there exists a vector $c \in T_{(p,x)}(P \times X)$ such that

$$Df(p, x)(c) = a.$$

Now,

$$T_{(p,x)}(P \times X) = T_p P \times T_x X,$$

so $c = (d, e)$ for some vectors $d \in T_p P$ and $e \in T_x X$. If d were zero, we would be done, for since the restriction of f to $\{p\} \times X$ is f_p, it follows that

$$a = Df(p, x)(0, e) = Df_p(x)(e).$$

Although d need not be zero, we may use the projection π to kill it off. As the derivative map

$$D\pi(p, x) \colon T_p P \times T_x X \to T_p P$$

is just projection onto the first factor, the regularity assumption that $D\pi(p, x)$ maps $T_{(p,x)}f^{-1}(y)$ onto $T_p P$ tells us that there is some vector of the form (d, v) in $T_{(p,x)}f^{-1}(y)$. But the restriction of f on the manifold $f^{-1}(y)$ is constant y, so

$$Df(p, x)(d, v) = 0.$$

Consequently, the vector $b := e - v \in T_x X$ is our solution because we have that

$$Df_p(x)(b) = Df(p,x)[(d,e) - (d,v)]$$
$$= Df(p,x)(d,e) - Df(p,x)(d,v)$$
$$= a - 0 = a.$$

$\square$

1.10 Curve selection lemma

Let S be a subset of $\mathbb{R}^n$ and $x^* \in \overline{S} \backslash S$. Then we can only say that there exists a sequence $\{x^k\}_k \subset S$ converging to x^* and, in general, there is no convergence rate estimates of the sequence to x^*. However, as shown by the next theorem, if the set S is semi-algebraic, we can find an analytic curve ϕ lying in S and satisfying

$$\|\phi(t) - x^*\| \le c\, t^\alpha \quad \text{for all } t \in [0, \epsilon)$$

for some positive constants c, α, and ϵ.

Theorem 1.11 (Curve Selection Lemma). *Let S be a semi-algebraic subset of $\mathbb{R}^n$, and $x^* \in \overline{S} \backslash S$. Then there exists a real analytic semi-algebraic curve*

$$\phi \colon (-\epsilon, \epsilon) \to \mathbb{R}^n$$

with $\phi(0) = x^*$ and with $\phi(t) \in S$ for $t \in (0, \epsilon)$.

Lemma 1.5 (Semi-algebraic choice). *Let S be a semi-algebraic subset of $\mathbb{R}^m \times \mathbb{R}^n$. Denote by $\pi \colon \mathbb{R}^m \times \mathbb{R}^n \to \mathbb{R}^m$ the projection on the first m coordinates. Then there is a semi-algebraic map $f \colon \pi(S) \to \mathbb{R}^n$ such that $(x, f(x)) \in S$ for all $x \in \pi(S)$.*

Proof. Since the natural projection $\mathbb{R}^m \times \mathbb{R}^n \to \mathbb{R}^m$ can be decomposed as

$$\mathbb{R}^m \times \mathbb{R}^n \to \mathbb{R}^m \times \mathbb{R}^{n-1} \to \cdots \to \mathbb{R}^m \times \mathbb{R} \to \mathbb{R}^m,$$

it is sufficient to consider the case $n = 1$. Furthermore, by using Theorem 1.1, we can reduce the proof of the lemma to the case where S is a cell of some cell decomposition of $\mathbb{R}^m \times \mathbb{R}$. Then $\pi(S)$ is a cell of $\mathbb{R}^m$. As a consequence, S is either the graph or cylinder over $\pi(S)$. If S is the graph of a semi-algebraic function $\xi \colon \pi(S) \to \mathbb{R}$, we define $f := \xi$. If S is a cylinder (ξ_1, ξ_2), we take

$$
f := \begin{cases}
\dfrac{1}{2}(\xi_1 + \xi_2) & \text{if } \xi_1 \neq -\infty \text{ and } \xi_2 \neq +\infty, \\[2mm]
\xi_2 - 1 & \text{if } \xi_1 = -\infty \text{ and } \xi_2 \neq +\infty, \\[2mm]
\xi_1 + 1 & \text{if } \xi_1 \neq -\infty \text{ and } \xi_2 = +\infty, \\[2mm]
0 & \text{if } \xi_1 = -\infty \text{ and } \xi_2 = +\infty.
\end{cases}
$$

Clearly, f has the desired properties. $\qquad\qquad\qquad\qquad\qquad\square$

In order to prove the Curve Selection Lemma, we recall Puiseux's theorem.

Lemma 1.6 (Puiseux's theorem). *Let*

$$
P(t, z) := a_0(t)z^d + a_1(t)z^{d-1} + \cdots + a_d(t),
$$

where $d \geq 1, a_0 \not\equiv 0$, and $a_0, \ldots, a_d$ are polynomials in one complex variable t. The following statements hold true:

(i) *For each $t \in \mathbb{C}$ with $a_0(t) \neq 0$, the polynomial $P(t, z)$ has exactly d (not necessarily distinct) complex roots $z = g_1(t), \ldots, z = g_d(t)$.*

(ii) *In a neighborhood of $0 \in \mathbb{C}$, each root g_i is a convergent fractional power series:*

$$
g_i(t) = \sum_{k \geq k_0} c_{i,k} t^{\frac{k}{m}} \quad \text{for } t \in \mathbb{C}, \ 0 < |t| < \epsilon,
$$

for some $\epsilon > 0, k_0, m \in \mathbb{Z}, m \geq 1$, and $c_{i,k} \in \mathbb{C}$. Furthermore, if there exists the finite limit $\lim_{t \to 0} g_i(t) \in \mathbb{C}$, then $c_{i,k} = 0$ for all $k < 0$ and so $t \mapsto g_i(t^m)$ is a complex analytic function in a neighborhood of 0.

Proof of Theorem 1.11. The map

$$
\| \cdot \| \colon \mathbb{R}^n \to \mathbb{R}, \quad x := (x_1, \ldots, x_n) \mapsto \sqrt{\sum_{i=1}^{n} x_i^2},
$$

is semi-algebraic. Since $x^* \in \overline{S} \backslash S$, the semi-algebraic set $\{\|x - x^*\| \mid x \in S\} \subset \mathbb{R}$ contains a sequence of positive real numbers tending to zero, hence it must contain an interval $(0, \epsilon)$ for some $\epsilon > 0$. Applying Lemma 1.5 to the semi-algebraic set $\{(\|x - x^*\|, x) \mid x \in S\} \subset \mathbb{R} \times \mathbb{R}^n$, we find a semi-algebraic map $f := (f_1, \ldots, f_n) \colon [0, \epsilon) \to \mathbb{R}^n$ such that $f(0) = x^*$, $f(t) \in S$ for all $t \in (0, \epsilon)$, and f is continuous at 0. By decreasing ϵ if necessary and using Theorem 1.7, we can see that each f_i is an analytic function on $(0, \epsilon)$.

Next, by Lemma 1.4, for each f_i, there exists a non-zero polynomial $P_i \colon \mathbb{R} \times \mathbb{R} \to \mathbb{R}$ such that $P_i(t, f_i(t)) = 0$ for all $t \in [0, \epsilon)$. It follows from Puiseux's theorem that there exist an integer $m_i \geq 1$ and a complex analytic function $g_i \colon \{t \in \mathbb{C} \mid |t| < \epsilon\} \to \mathbb{C}$ (after perhaps reducing ϵ) such that $f_i(t^{m_i}) = g_i(t)$ for $t \in [0, \epsilon)$.

Finally, we leave to the reader to verify that the curve

$$\phi(t) := (g_1(t^{m_2 \cdots m_n}), \ldots, g_n(t^{m_1 \cdots m_{n-1}})) \quad \text{for } t \in (-\epsilon, \epsilon)$$

has the desired properties. $\qquad \square$

Theorem 1.12 (Curve Selection Lemma at infinity). *Let* $S \subset \mathbb{R}^n$ *be a semi-algebraic set, and let*

$$f := (f_1, \ldots, f_p) \colon \mathbb{R}^n \to \mathbb{R}^p$$

be a semi-algebraic map. Assume that there exists a sequence $\{x^k\}$ *such that* $x^k \in S$, $\lim_{k \to \infty} \|x^k\| = \infty$ *and* $\lim_{k \to \infty} f(x^k) = y \in (\overline{\mathbb{R}})^p$, *where* $\overline{\mathbb{R}} := \mathbb{R} \cup \{\pm\infty\}$. *Then there exists a meromorphic semi-algebraic curve*

$$\phi \colon (0, \epsilon) \to \mathbb{R}^n$$

such that $\phi(t) \in S$ *for all* $t \in (0, \epsilon), \lim_{t \to 0} \|\phi(t)\| = \infty$, *and* $\lim_{t \to 0} f(\phi(t)) = y$.

Proof. By replacing, if necessary, f_i by $\frac{\pm 1}{1 \pm (f_i(x))^2}$, we may assume that $y \in \mathbb{R}^p$.

Consider the semi-algebraic map $\Phi \colon \mathbb{R}^n \to \mathbb{R}^{n+1} \times \mathbb{R}^p$ given by

$$\Phi(x) := \left(\frac{x_1}{\sqrt{1 + \|x\|^2}}, \ldots, \frac{x_n}{\sqrt{1 + \|x\|^2}}, \frac{1}{\sqrt{1 + \|x\|^2}}, f(x) \right).$$

Without loss of generality, we can suppose that the sequence $\Phi(x^k)$ is convergent to some point $(u, y) \in \mathbb{S}^n \times \mathbb{R}^p$. By the Tarski–Seidenberg theorem, $\Phi(S)$ is a semi-algebraic set. Then by Curve Selection Lemma (Theorem 1.11), there exists an analytic semi-algebraic curve

$$\psi(t) \colon (-\epsilon, \epsilon) \to \mathbb{R}^{n+1} \times \mathbb{R}^p,$$

$$t \mapsto (\psi_1(t), \ldots, \psi_n(t), \psi_{n+1}(t), \ldots, \psi_{n+1+p}(t)),$$

such that

$$\psi(0) = (u, y) \quad \text{and} \quad \psi(t) \in \Phi(S) \text{ for all } t \in (0, \epsilon).$$

Now, define the curve $\phi \colon (0, \epsilon) \to \mathbb{R}^n, t \mapsto \phi(t)$, by

$$\phi(t) := \left(\frac{\psi_1(t)}{\psi_{n+1}(t)}, \ldots, \frac{\psi_n(t)}{\psi_{n+1}(t)} \right).$$

Then it is clear that ϕ has the required properties. □

1.11   Growth dichotomy lemma

Lemma 1.7 (Growth Dichotomy Lemma). (i) *Let* $f \colon (0, \epsilon) \to \mathbb{R}$ *be a semi-algebraic function with* $f(t) \neq 0$ *for all* $t \in (0, \epsilon)$. *Then there exist constants* $a \neq 0$ *and* $\alpha \in \mathbb{Q}$ *such that* $f(t) = at^\alpha + o(t^\alpha)$ *as* $t \to 0^+$.

(ii) *Let* $f \colon (r, +\infty) \to \mathbb{R}$ *be a semi-algebraic function with* $f(t) \neq 0$ *for all* $t \in (r, +\infty)$. *Then there exist constants* $a \neq 0$ *and* $\alpha \in \mathbb{Q}$ *such that* $f(t) = at^\alpha + o(t^\alpha)$ *as* $t \to +\infty$.

Proof. This is a direct consequence of Theorems 1.11 and 1.12. □

1.12   Connected components of semi-algebraic sets

Theorem 1.13. *The following statements hold.*

(i) *Every semi-algebraic set has a finite number of connected components and each such component is semi-algebraic.*

(ii) *Every semi-algebraic set* S *is locally connected, i.e. for every* $x \in S$, *every ball* B *with center* x *contains a connected neighborhood of* x *in* S.

(iii) *Every connected semi-algebraic set S is semi-algebraically arc-wise connected: for every points x, y in S, there exists a piecewise analytic semi-algebraic map*

$$\gamma \colon [0, 1] \to S$$

such that $\gamma(0) = x$ and $\gamma(1) = y$.

Proof. (i) The statement (i) is Theorem 1.2.

(ii) Let $x \in S$ and B an open ball, $x \in B$. Then $B \cap S$ is semi-algebraic, hence it has a finite number of connected components. Let C be the connected component containing x. Then C is a connected neighborhood of x in S.

(iii) By Theorem 1.3, we can write $S = \bigcup_{i=1}^{p} C_i$, where C_i is semi-algebraically homeomorphic to an open hypercube $(0, 1)^{d_i} \subset \mathbb{R}^{d_i}$. Since S is connected, we can number $C_1, \ldots, C_p$ in such a way that $C_i \cap \overline{C}_{i+1} \neq \emptyset$. Now, the statement (iii) follows from the Curve Selection Lemma (Theorem 1.11) and Theorem 1.8. $\qquad\square$

1.13 Łojasiewicz's inequalities

The following result gives a relative rate of growth of two continuous semi-algebraic functions.

Theorem 1.14 (Łojasiewicz's inequality). *Let K be a compact semi-algebraic subset of $\mathbb{R}^n$. Let $f, g \colon K \to \mathbb{R}$ be continuous semi-algebraic functions such that $g^{-1}(0) \subset f^{-1}(0)$. Then there exist $c > 0$ and $\alpha > 0$ such that*

$$|g(x)| \geq c|f(x)|^{\alpha} \quad \text{for all } x \in K.$$

Proof. We assume that $g^{-1}(0)$ is non-empty and different from K, otherwise the theorem is trivial.

Let $M := \max_{x \in K} |f(x)|$ and consider the function

$$\mu \colon [0, M] \to \mathbb{R}, \quad t \mapsto \mu(t) := \inf\{|g(x)| \mid x \in K, |f(x)| = t\}.$$

We can see that μ is semi-algebraic, continuous at 0, $\mu(0) = 0$, and $\mu(t) \neq 0$ for $0 < t \ll 1$. Thanks to Growth Dichotomy Lemma

(Lemma 1.7(i)), we can write

$$\mu(t) = at^\alpha + o(t^\alpha) \quad \text{as } t \to 0^+,$$

for some constants $a > 0$ and $\alpha > 0$. Consequently, there are $c > 0$ and $0 < \delta \leq M$ such that

$$\mu(t) \geq ct^\alpha \quad \text{for all } t \in [0, \delta].$$

Now take any $x \in K$ with $t := |f(x)| \leq \delta$. We have

$$|g(x)| \geq \mu(t) \geq ct^\alpha = c|f(x)|^\alpha.$$

Since $g^{-1}(0) \subset f^{-1}(0)$, the inequality

$$|g(x)| \geq c|f(x)|^\alpha$$

still holds for all $x \in K$ with $|f(x)| \geq \delta$, after perhaps reducing c. $\qquad\square$

Corollary 1.2 (The classical Łojasiewicz inequality). *Let K be a semi-algebraic compact subset of $\mathbb{R}^n$ and $f \colon K \to \mathbb{R}$ be a continuous semi-algebraic function. Then there exist $c > 0$ and $\alpha > 0$ such that*

$$|f(x)| \geq c\,\mathrm{dist}(x, f^{-1}(0))^\alpha \quad \text{for all } x \in K.$$

Proof. This is a direct consequence of Theorem 1.14. $\qquad\square$

The next result will be useful in the proof of Łojasiewicz's gradient inequality below.

Lemma 1.8 (Bochnack–Łojasiewicz's inequality). *Let f be a semi-algebraic function of class C^1 in a neighborhood of $x^* \in \mathbb{R}^n$ and let $c \in (0, 1)$. Then*

$$\|x - x^*\|\,\|\nabla f(x)\| \geq c\,|f(x) - f(x^*)|$$

in a neighborhood of x^*.

Proof. Without loss of generality, we may assume that $x^* = 0$ and $f(x^*) = 0$.

We will prove the lemma by contradiction. Suppose that 0 is in the closure of the set $\{x \mid \|x\|\,\|\nabla f(x)\| < c|f(x)|\}$; this set is semi-algebraic, hence (by Curve Selection Lemma (Theorem 1.11)) it contains an arc $\{x = \phi(t) \mid 0 < t < \epsilon\}$, where ϕ is C^1 in $[0, \epsilon)$, $\phi(0) = 0$, and $\|\phi'(t)\| = 1$. This implies, putting $g(t) := f(\phi(t))$, that $|t| \|g'(t)\| \leq \rho |g(t)|$ in $[0, \delta]$ with some $c < \rho < 1$ and $0 < \delta < \epsilon$. This gives easily $|g(t)| \geq (t/\delta)^\rho |g(\delta)|$, which is impossible. □

The following result gives information concerning the relative rate of growth of semi-algebraic functions and their gradient norms.

Theorem 1.15 (Łojasiewicz's gradient inequality). *Let f be a semi-algebraic function of class C^1 in a neighborhood of $0 \in \mathbb{R}^n$ such that $f(0) = 0$. Then there exist constants $c > 0$ and $\alpha \in [0, 1)$ such that, for all x in a neighborhood of 0,*

$$\|\nabla f(x)\| \geq c |f(x)|^\alpha.$$

Proof. We assume that $\nabla f(0) = 0$, otherwise the theorem is trivial. By applying Lemma 1.8 with $x^* = 0$, there is a compact ball K centered at 0 in which $\nabla f(x) = 0$ only if $f(x) = 0$. As in the proof of Theorem 1.14, let

$$\mu(t) := \inf\{\|\nabla f(x)\| \mid x \in K, |f(x)| = t\}.$$

Then we can write

$$\mu(t) = at^\alpha + o(t^\alpha) \quad \text{as } t \to 0^+$$

for some constants $a > 0$ and $\alpha > 0$, and hence there exists a constant $c > 0$ such that

$$\|\nabla f(x)\| \geq c |f(x)|^\alpha \quad \text{for all } x \in K.$$

Furthermore, by the definition of μ, there exists a semi-algebraic function $\phi \colon [0, \epsilon) \to K$ such that, for all $t \in [0, \epsilon)$,

$$\|\nabla f(\phi(t))\| = \mu(t) \quad \text{and} \quad |f(\phi(t))| = t.$$

Consequently, we obtain

$$\|\nabla f(\phi(t))\| = a |f(\phi(t))|^\alpha + o(t^\alpha) \quad \text{as } t \to 0^+.$$

Since K is compact, there exists the limit $x^* := \lim_{t \to 0^+} \phi(t) \in K$. It is clear that

$$\|\nabla f(x^*)\| = \mu(0) = 0 \quad \text{and} \quad |f(x^*)| = 0.$$

According to Lemma 1.8 again, we have for some $c' \in (0,1)$ and for all t in a neighborhood of 0,

$$c'|f(\phi(t))| \leq \|\phi(t) - x^*\| \|\nabla f(\phi(t))\|$$
$$= a\|\phi(t) - x^*\| |f(\phi(t))|^\alpha + o(t^\alpha),$$

which implies easily $\alpha < 1$. $\square$

Finally, we state, without proof, an effective version of Łojasiewicz's gradient inequality for polynomials.

Theorem 1.16. *Let* $f \colon \mathbb{R}^n \to \mathbb{R}$ *be a polynomial of degree* $d \geq 1$. *Suppose that* $f(0) = 0$ *and* $\nabla f(0) = 0$. *Then there exist a neighborhood* U *of* 0 *and a constant* $c > 0$ *such that for all* $x \in U$, *we have*

$$\|\nabla f(x)\| \geq c|f(x)|^{1 - \frac{1}{\mathscr{R}(n,d)}},$$

where

$$\mathscr{R}(n,d) := \begin{cases} d(3d-3)^{n-1} & \text{if } d \geq 2, \\ 1 & \text{if } d = 1. \end{cases}$$

Bibliographic notes

The main references on semi-algebraic geometry are the texts in [15, 22]. In this chapter, as in [93], we choose the following logical order in representing results:

$$\text{Cell decomposition} \Rightarrow \text{Tarski–Seidenberg theorem}$$
$$\Rightarrow \text{other results.}$$

Lemmas 1.1–1.3 were also taken from [93]. Lemma 1.4 and its proof are given in [15]. The proof of Theorem 1.1 was adapted from that of the Cylindrical Decomposition Theorem of [93]. The proof of Theorem 1.3 can be found in [22, Proposition 3.7 and Corollary 3.8]. The proof of Theorem 1.7 was suggested by Tạ Lê Lợi (private communication). Theorem 1.8 and its proof are given in [34], where the author considered the (more general) case of a function defined in some o-minimal structure. The proof of Theorem 1.10 is essentially the same as that of [45, The Transversality Theorem]. The statement (but not the proof) of this version of the Curve Selection Lemma (Theorem 1.11) is taken from [101]. Theorem 1.12 can be found in the papers in [28, 106]. Puiseux's theorem and its proof can be found in [20]. The proof of Theorem 1.14 matches closely with that of [14, Théorème 1]. Lemma 1.8 was taken from [13]. Theorem 1.16 and its proof can be found in [1].

Chapter 2

Critical Points and Tangencies

Abstract

Given a semi-algebraic C^1-function $f \colon \mathbb{R}^n \to \mathbb{R}$ and a basic closed semi-algebraic set $S \subset \mathbb{R}^n$, we introduce and establish some properties of the set $\Sigma(f, S)$ of critical points of f on S and of the tangency variety $\Gamma(f, S)$ of f on S. These sets are closely related to the optimization problem $\inf_{x \in S} f(x)$, and will be encountered later in Chapter 8.

2.1 Regularity

In this section, we recall briefly the definition of regularity and the first-order necessary conditions for constrained optimality.

2.1.1 *Optimality conditions*

Let $f, g_i, h_j \colon \mathbb{R}^n \to \mathbb{R}$, $i = 1, \ldots, l, j = 1, \ldots, m$, be C^1-functions and assume that the set

$$S := \{x \in \mathbb{R}^n \,|\, g_1(x) = 0, \ldots, g_l(x) = 0, \ h_1(x) \geq 0, \ldots, h_m(x) \geq 0\}$$

is non-empty. Consider the optimization problem

$$\text{minimize } f(x) \quad \text{for all } x \in S.$$

The standard first-order necessary conditions for local optimality in this problem are the following.

Theorem 2.1 (Fritz-John optimality conditions). *If* $x \in S$ *is a local minimizer of* f *on* S, *then there exist real numbers* κ, λ_i, $i = 1, \ldots, l$, *and* $\nu_j, j = 1, \ldots, m$, *not all zero, such that*

$$\kappa \nabla f(x) - \sum_{i=1}^{l} \lambda_i \nabla g_i(x) - \sum_{j=1}^{m} \nu_j \nabla h_j(x) = 0,$$

$$\nu_j h_j(x) = 0, \; \nu_j \geq 0, \quad \text{for } j = 1, \ldots, m.$$

Note that if $\kappa = 0$, the above conditions are not very informative about a local minimizer and so, we usually make an assumption called a constraint qualification to ensure that $\kappa \neq 0$. A constraint qualification — probably the one most often used in the design of algorithms — is defined as follows.

Definition 2.1. The constraint set S is said to be *regular* at $x \in S$ if the gradient vectors $\nabla g_i(x)$, $i = 1, \ldots, l$, and $\nabla h_j(x)$, $j \in J(x)$, are linearly independent, where

$$J(x) := \{j \in \{1, \ldots, m\} \mid h_j(x) = 0\}$$

is called the set of *active constraint indices*. S is called *regular* if it is regular at every point $x \in S$.

Under the assumption that S is regular, we may obtain the more informative optimality conditions due to Karush, Kuhn and Tucker (and called the KKT optimality conditions) where the real number κ in Theorem 2.1 can be taken to be 1.

Theorem 2.2 (KKT optimality conditions). *Let* $x \in S$ *be a local minimizer of* f *on* S. *If* S *is regular at* x, *then there exist real numbers* $\lambda_i, i = 1, \ldots, l$, *and* $\nu_j, j = 1, \ldots, m$, *such that*

$$\nabla f(x) - \sum_{i=1}^{l} \lambda_i \nabla g_i(x) - \sum_{j=1}^{m} \nu_j \nabla h_j(x) = 0,$$

$$\nu_j h_j(x) = 0, \; \nu_j \geq 0, \quad \text{for } j = 1, \ldots, m.$$

Proof. This is a direct consequence of Theorem 2.1. □

The real numbers λ_i and ν_j are called the *Lagrange multipliers*.

2.1.2 *Transversality to large spheres*

The following lemma says that if the constraint set S is unbounded and regular, then it intersects transversally with the sphere $\{x \in \mathbb{R}^n \mid \|x\|^2 = R^2\}$ for all R large enough.

Lemma 2.1. *Suppose the set S is unbounded and regular. Then there exists a real number $R_0 > 0$ such that for all $R \geq R_0$, the set*

$$S_R := \{x \in S \mid \|x\|^2 = R^2\}$$

is a non-empty compact semi-algebraic set, and it is regular, i.e. for each $x \in S_R$, the vectors x, $\nabla g_i(x), i = 1, \ldots, l$, and $\nabla h_j(x), j \in J(x)$, are linearly independent.

Proof. We first prove that $S_R \neq \emptyset$ for all R large enough. Indeed, since S is unbounded, there exists a sequence $x^k \in S$ with $\|x^k\| \to +\infty$. Thanks to the Curve Selection Lemma at infinity (Theorem 1.12), we can find a smooth semi-algebraic curve $\varphi \colon (0, \epsilon] \to \mathbb{R}^n, t \mapsto \varphi(t)$, such that $\varphi(t) \in S$ for all $t \in (0, \epsilon]$ and $\|\varphi(t)\| \to +\infty$ as $t \to 0^+$. By the continuity of φ, we have for each $R \geq \|\varphi(\epsilon)\|$, there exists $t \in (0, \epsilon]$ such that $R = \|\varphi(t)\|$. In consequence, $S_R \neq \emptyset$ for all $R \geq \|\varphi(\epsilon)\|$.

We next show that the set S_R is regular for R large enough. By contradiction, assume that there exist sequences $x^k \in S$, with $\|x^k\| \to +\infty$, and $\lambda_i^k, \nu_j^k, \mu_k \in \mathbb{R}$, not all zero, for $k \in \mathbb{N}$ such that

$$\sum_{i=1}^l \lambda_i^k \nabla g_i(x^k) + \sum_{j \in J(x^k)} \nu_j^k \nabla h_j(x^k) + \mu_k x^k = 0.$$

Since the number of subsets of $\{1, \ldots, m\}$ is finite, we may assume without loss of generality that $J(x^k) = J$ for all k and some subset

J of $\{1, \ldots, m\}$. Then the set

$$\Bigg\{ x \in S \,|\, \text{there exist } \lambda_i, \nu_j, \mu \in \mathbb{R}, \text{ not all zero, such that}$$

$$\sum_{i=1}^{l} \lambda_i g_i(x) + \sum_{j \in J} \nu_j \nabla h_j(x) + \mu x = 0,$$

$$h_j(x) = 0, \text{ for } j \in J \Bigg\}$$

is unbounded and semi-algebraic (because of the Tarski–Seidenberg theorem (Theorem 1.5). Using the Curve Selection Lemma at infinity (Theorem 1.12), there exist a smooth semi-algebraic curve $\varphi(t)$ and semi-algebraic functions $\lambda_i(t), \nu_j(t), \mu(t), t \in (0, \epsilon]$, not all zero, such that

(a) $\varphi(t) \in S$ for $t \in (0, \epsilon]$;
(b) $\|\varphi(t)\| \to +\infty$ as $t \to 0^+$;
(c) $\sum_{i=1}^{l} \lambda_i(t) \nabla g_i(\varphi(t)) + \sum_{j \in J} \nu_j(t) \nabla h_j(\varphi(t)) + \mu(t)\varphi(t) \equiv 0$; and
(d) $h_j(\varphi(t)) \equiv 0$ for all $j \in J$.

Hence, for all $t \in (0, \epsilon]$,

$$0 = \sum_{i=1}^{l} \lambda_i(t) \left\langle \nabla g_i(\varphi(t)), \frac{d\varphi}{dt} \right\rangle + \sum_{j \in J} \nu_j(t) \left\langle \nabla h_j(\varphi(t)), \frac{d\varphi}{dt} \right\rangle$$

$$+ \mu(t) \left\langle \varphi(t), \frac{d\varphi}{dt} \right\rangle$$

$$= \sum_{i=1}^{l} \lambda_i(t) \frac{d}{dt}(g_i \circ \varphi)(t) + \sum_{j \in J} \nu_j(t) \frac{d}{dt}(h_j \circ \varphi)(t) + \frac{\mu(t)}{2} \frac{d\|\varphi(t)\|^2}{dt}$$

$$= \frac{\mu(t)}{2} \frac{d\|\varphi(t)\|^2}{dt}.$$

So $\mu(t) \equiv 0$, which contradicts the regularity of the set S. $\qquad\square$

2.2 Critical points

Let $f, g_i, h_j \colon \mathbb{R}^n \to \mathbb{R}$, $i = 1, \ldots, l$, $j = 1, \ldots, m$, be semi-algebraic C^1-functions and assume that the set

$$S := \{x \in \mathbb{R}^n \mid g_1(x) = 0, \ldots, g_l(x) = 0, \ h_1(x) \geq 0, \ldots, h_m(x) \geq 0\}$$

is non-empty.

Definition 2.2. We define the *set of critical points of f on S* to be the set:

$$\Sigma(f, S) := \left\{ x \in S \mid \text{there exist } \lambda_i, \nu_j \in \mathbb{R} \text{ such that} \right.$$

$$\nabla f(x) - \sum_{i=1}^{l} \lambda_i \nabla g_i(x) - \sum_{j=1}^{m} \nu_j \nabla h_j(x) = 0,$$

$$\left. \nu_j h_j(x) = 0, j = 1, \ldots, m \right\}.$$

Let $K_0(f, S) := f(\Sigma(f, S))$ and we call it the *set of critical values of f on S*.

In the case $S = \mathbb{R}^n$, for simplicity, we will write $\Sigma(f)$ and $K_0(f)$ instead of $\Sigma(f, S)$ and $K_0(f, S)$, respectively. By definition, then

$$\Sigma(f) = \{x \in \mathbb{R}^n \mid \nabla f(x) = 0\},$$

which is the usual *set of critical points* of f.

Remark 2.1. It is worth noting that we do not include the condition that the Lagrange multipliers ν_j are non-negative, as is usual. It turns out that we do not need the non-negativeness of ν_j to obtain the representation theorems in Section 8.3, as we shall see. Since taking the sign of ν_j into account adds an unnecessary complication to the representation theorems, we omit it.

We will give some properties of the sets $\Sigma(f, S)$ and $K_0(f, S)$. We first have the following lemma.

Lemma 2.2. *The following statements hold true:*

(i) $\Sigma(f, S) = \Sigma(f + c, S)$ *for all* $c \in \mathbb{R}$.

(ii) $\Sigma(f, S)$ *is a semi-algebraic (possibly empty) set.*

(iii) *Let* $x \in S$ *be a minimizer of* f *on* S. *If* S *is regular at* x, *then* $x \in \Sigma(f, S)$.

Proof. (i) This is straightforward.

(ii) Let

$$A := \left\{ (x, \lambda, \nu) \in \mathbb{R}^n \times \mathbb{R}^l \times \mathbb{R}^m \mid \nabla f(x) - \sum_{i=1}^{l} \lambda_i \nabla g_i(x) \right.$$

$$\left. - \sum_{j=1}^{m} \nu_j \nabla h_j(x) = 0, \quad g_i(x) = 0, i = 1, \ldots, l, \right.$$

$$\left. h_j(x) \geq 0 \text{ and } \nu_j h_j(x) = 0, j = 1, \ldots, m \right\}.$$

Clearly, A is a semi-algebraic set and $\Sigma(f, S) = \pi(A)$, where $\pi(x, \lambda, \nu) := x$-the projection on the first component. By the Tarski–Seidenberg theorem (Theorem 1.5), the set $\Sigma(f, S)$ is semi-algebraic.

(iii) This is a direct consequence of the KKT optimality conditions. $\qquad \square$

The following result, which will be used in Chapter 8, is a special case of Sard's theorem.

Theorem 2.3. *The set* $K_0(f, S)$ *of critical values of* f *on* S *is finite.*

Proof. Let A be the following semi-algebraic set

$$\left\{ (x, \lambda, \nu) \in \mathbb{R}^n \times \mathbb{R}^l \times \mathbb{R}^m \mid \nabla f(x) - \sum_{i=1}^{l} \lambda_i \nabla g_i(x) \right.$$

$$\left. - \sum_{j=1}^{m} \nu_j \nabla h_j(x) = 0, \quad g_i(x) = 0, i = 1, \ldots, l, \right.$$

$$\left. h_j(x) \geq 0 \text{ and } \nu_j h_j(x) = 0, j = 1, \ldots, m \right\}.$$

We have $\Sigma(f,S) = \pi(A)$, where $\pi(x,\lambda,\nu) := x$ — the projection on the first component. It suffices to show that $(f \circ \pi)(A)$ is a finite set.

Indeed, the set A is semi-algebraic, and so it has finitely many connected components, each of which being semi-algebraic. Then we may write

$$A = A_1 \cup A_2 \cup \cdots \cup A_s,$$

where each A_k is a connected semi-algebraic set.

Let $(\varphi, \lambda, \nu) \colon [0,1] \to \mathbb{R}^n \times \mathbb{R}^l \times \mathbb{R}^m$ be a smooth semi-algebraic curve such that $(\varphi(t), \lambda(t), \nu(t)) \in A_k$ for all $t \in [0,1]$. We have

(a) $\nabla f(\varphi(t)) - \sum_{i=1}^{l} \lambda_i(t) \nabla g_i(\varphi(t)) - \sum_{j=1}^{m} \nu_j(t) \nabla h_j(\varphi(t)) \equiv 0$;

(b) $g_i(\varphi(t)) \equiv 0, i = 1, \ldots, l$; and

(c) $\nu_j(t) h_j(\varphi(t)) \equiv 0, j = 1, \ldots, m$.

Since the functions ν_j and $h_j \circ \varphi$ are semi-algebraic, it follows from the Monotonicity Theorem (Theorem 1.8) that there is a partition $0 =: t_1 < \cdots < t_N := 1$ of $[0,1]$ such that on each interval (t_l, t_{l+1}), these functions are either constant or strictly monotone, for $l \in \{1, \ldots, N-1\}$. Then, by (c), we can see that either $\nu_j(t) \equiv 0$ or $(h_j \circ \varphi)(t) \equiv 0$ on (t_l, t_{l+1}). In particular, we have for all $t \in [0,1]$,

$$\nu_j(t) \frac{d}{dt}(h_j \circ \varphi)(t) \equiv 0, \quad j = 1, \ldots, m.$$

It follows from (a) and (b) that

$$
\begin{aligned}
\frac{d}{dt}(f \circ \varphi)(t) &= \left\langle \nabla f(\varphi(t)), \frac{d\varphi(t)}{dt} \right\rangle \\
&= \sum_{i=1}^{l} \lambda_i(t) \left\langle \nabla g_i(\varphi(t)), \frac{d\varphi(t)}{dt} \right\rangle \\
&\quad + \sum_{j=1}^{m} \nu_j(t) \left\langle \nabla h_j(\varphi(t)), \frac{d\varphi(t)}{dt} \right\rangle \\
&= \sum_{i=1}^{l} \lambda_i(t) \frac{d}{dt}(g_i \circ \varphi)(t) + \sum_{j=1}^{m} \nu_j(t) \frac{d}{dt}(h_j \circ \varphi)(t) \\
&= 0.
\end{aligned}
$$

So f is constant on the curve φ. This implies easily that $f \circ \pi$ is constant on the curve (φ, λ, ν).

On the other hand, since the semi-algebraic set A_k is connected, any two points in A_k can be joined (in A_k) by a piecewise smooth semi-algebraic curve (see Theorem 1.13). It follows that $f \circ \pi$ is constant on A_k. Therefore, the image $(f \circ \pi)(A_k)$ is a finite set. This completes the proof.   □

2.3   Tangencies

Let $f, g_i, h_j \colon \mathbb{R}^n \to \mathbb{R}$, $i = 1, \ldots, l, j = 1, \ldots, m$, be semi-algebraic C^1-functions, and assume that the set

$$S := \{x \in \mathbb{R}^n \mid g_1(x) = 0, \ldots, g_l(x) = 0, \ h_1(x) \geq 0, \ldots, h_m(x) \geq 0\}$$

is non-empty.

2.3.1   *Tangency varieties*

Let us start with the following definition.

Definition 2.3. By the *tangency variety of f on S* we mean the set

$$\Gamma(f, S) := \left\{ x \in S \mid \text{there exist } \kappa, \lambda_i, \nu_j, \mu \in \mathbb{R}, \text{ not all zero, such that} \right.$$

$$\kappa \nabla f(x) - \sum_{i=1}^{l} \lambda_i \nabla g_i(x) - \sum_{j=1}^{m} \nu_j \nabla h_j(x) - \mu x = 0, \text{ and}$$

$$\left. \nu_j h_j(x) = 0, j = 1, \ldots, m \right\}.$$

In the case $S = \mathbb{R}^n$, for simplicity, we will write $\Gamma(f)$ instead of $\Gamma(f, S)$. By definition, then

$$\Gamma(f) = \left\{ x \in \mathbb{R}^n \middle| \text{rank} \begin{pmatrix} \dfrac{\partial f}{\partial x_1} & \cdots & \dfrac{\partial f}{\partial x_n} \\ x_1 & \cdots & x_n \end{pmatrix} \leq 1 \right\}.$$

Geometrically, the tangency variety $\Gamma(f,S)$ of f consists of all points $x \in S$ where the level sets of the restriction $f|_S$ of f on S are tangent to the sphere in $\mathbb{R}^n$ centered in the origin with radius $\|x\|$.

Remark 2.2. (i) We can replace the tangency variety $\Gamma(f,S)$ by the set

$$\Gamma_a(f,S) := \left\{ x \in S \,\middle|\, \text{there exist } \kappa, \lambda_i, \nu_j, \mu, \text{ not all zero, such that} \right.$$

$$\kappa \nabla f(x) - \sum_{i=1}^{l} \lambda_i \nabla g_i(x) - \sum_{j=1}^{m} \nu_j \nabla h_j(x) - \mu(x - a) = 0,$$

$$\left. \nu_j h_j(x) = 0, j = 1, \ldots, m \right\},$$

where $a \in \mathbb{R}^n$. Then all subsequent results will still hold with obvious modifications. The advantage is that if the center a is general enough, then $\Gamma_a(f,S) \setminus \Sigma(f,S)$ is a one-dimensional submanifold of $\mathbb{R}^n$.

(ii) As in the definition of the set $\Sigma(f,S)$, we do not include the condition that the Lagrange multipliers ν_j are non-negative. It turns out that we do not need the non-negativeness of ν_j to obtain the representation theorems in Section 8.4. Since taking the sign of ν_j into account adds an unnecessary complication to the representation theorems, we omit it.

In this subsection, we will establish some properties of the tangency variety $\Gamma(f,S)$. We start with the following simple lemma.

Lemma 2.3. *The following statements hold true:*

(i) $\Gamma(f,S) = \Gamma(f+c,S)$ *for all* $c \in \mathbb{R}$.
(ii) $\Sigma(f,S) \subset \Gamma(f,S)$.
(iii) $\Gamma(f,S)$ *is a non-empty semi-algebraic set.*
(iv) *If S is unbounded, so is $\Gamma(f,S)$.*

Proof. (i) and (ii) are straightforward.

(iii)–(iv) By the Tarski–Seidenberg theorem (Theorem 1.5), it is easy to check that the set $\Gamma(f,S)$ is semi-algebraic.

We next show that $\Gamma(f, S) \neq \emptyset$. To this end, take any $a \in S$. Then the set $S_a := \{x \in S \mid \|x\|^2 = \|a\|^2\}$ is non-empty compact. Hence, the optimization problem $\min_{x \in S_a} f(x)$ has a global minimizer, say $b \in S_a$. The Fritz-John optimality conditions (Theorem 2.1) imply that there exist real numbers κ, λ_i, ν_j, and μ, not all zero, such that

$$\kappa \nabla f(b) - \sum_{i=1}^{l} \lambda_i \nabla g_i(b) - \sum_{j=1}^{m} \nu_j \nabla h_j(b) - \mu b = 0, \quad \text{and}$$

$$\nu_j h_j(b) = 0, \quad j = 1, \ldots, m.$$

By definition, then $b \in \Gamma(f, S)$, and so $\Gamma(f, S) \neq \emptyset$.

Finally, it is clear that if $\|a\| \to \infty$ then $\|b\| = \|a\| \to \infty$, which proves (iv). □

Another property of the tangency variety $\Gamma(f, S)$ is stated in the following lemma.

Lemma 2.4. *If the set S is unbounded and regular, then for each $x \in \Gamma(f, S), \|x\| \gg 1$, there exist real numbers λ_i, ν_j, and μ such that*

$$\nabla f(x) - \sum_{i=1}^{l} \lambda_i \nabla g_i(x) - \sum_{j=1}^{m} \nu_j \nabla h_j(x) - \mu x = 0, \quad \text{and}$$

$$\nu_j h_j(x) = 0, \quad j = 1, \ldots, m.$$

Proof. Since S is unbounded, so is $\Gamma(f, S)$. Let $x \in \Gamma(f, S)$. By definition, there exist real numbers $\kappa, \lambda_i, \nu_j, \mu$, at least one of which is different from zero, such that

$$\kappa \nabla f(x) - \sum_{i=1}^{l} \lambda_i \nabla g_i(x) - \sum_{j=1}^{m} \nu_j \nabla h_j(x) - \mu x = 0,$$

$$\nu_j h_j(x) = 0, \quad j = 1, \ldots, m.$$

Thus, it suffices to show that $\kappa \neq 0$, provided that $x \in \Gamma(f, S)$, $\|x\| \gg 1$.

By contradiction and using the Curve Selection Lemma at infinity (Theorem 1.12), there exist a smooth semi-algebraic curve $\varphi(t)$ and semi-algebraic functions $\lambda_i(t), \nu_j(t), \mu(t), t \in (0, \epsilon]$, such that

(a) $\varphi(t) \in S$ for $t \in (0, \epsilon]$;
(b) $\|\varphi(t)\| \to +\infty$ as $t \to 0^+$;
(c) $\sum_{i=1}^l \lambda_i(t)\nabla g_i(\varphi(t)) + \sum_{j=1}^m \nu_j(t)\nabla h_j(\varphi(t)) + \mu(t)\varphi(t) \equiv 0$; and
(d) $\nu_j(t)h_j(\varphi(t)) \equiv 0, j = 1, \ldots, m$.

Since the functions ν_j and $h_j \circ \varphi$ are semi-algebraic, it follows from the Monotonicity Theorem (Theorem 1.8) that for $\epsilon > 0$ small enough, these functions are either constant or strictly monotone. Then, by (d), we can see that either $\nu_j(t) \equiv 0$ or $(h_j \circ \varphi)(t) \equiv 0$; in particular,

$$\nu_j(t)\frac{d}{dt}(h_j \circ \varphi)(t) \equiv 0, \quad j = 1, \ldots, m.$$

Hence, it follows from (c) that

$$0 = \sum_{i=1}^l \lambda_i(t) \left\langle \nabla g_i(\varphi(t)), \frac{d\varphi}{dt} \right\rangle + \sum_{j=1}^m \nu_j(t) \left\langle \nabla h_j(\varphi(t)), \frac{d\varphi}{dt} \right\rangle$$

$$+ \mu(t) \left\langle \varphi(t), \frac{d\varphi}{dt} \right\rangle$$

$$= \sum_{i=1}^l \lambda_i(t)\frac{d}{dt}(g_i \circ \varphi)(t) + \sum_{j=1}^m \nu_j(t)\frac{d}{dt}(h_j \circ \varphi)(t) + \frac{\mu(t)}{2}\frac{d\|\varphi(t)\|^2}{dt}$$

$$= \frac{\mu(t)}{2}\frac{d\|\varphi(t)\|^2}{dt}.$$

So $\mu(t) \equiv 0$, which contradicts the regularity of the set S. □

The next lemma will be used in the proof of Theorem 2.6 below.

Lemma 2.5. *Assume that the set S is regular and let $c \in \mathbb{R}$. If the set*

$$\{x \in S \mid f(x) = c \text{ and } x \in \Gamma(f, S)\}$$

is unbounded, then c is a critical value of f on S, i.e. $c \in K_0(f, S)$.

Proof. By Lemma 2.4 and the Curve Selection Lemma at infinity (Theorem 1.12), there exist a smooth semi-algebraic curve $\varphi(t)$ and semi-algebraic functions $\lambda_i(t), \nu_j(t), \mu(t)$, for $t \in (0, \epsilon]$, such that

(a) $\varphi(t) \in S$ for $t \in (0, \epsilon]$;
(b) $\|\varphi(t)\| \to +\infty$ as $t \to 0^+$;
(c) $f \circ \varphi(t) \equiv c$;
(d) $\nabla f(\varphi(t)) - \sum_{i=1}^{l} \lambda_i(t) \nabla g_i(\varphi(t)) - \sum_{j=1}^{m} \nu_j(t) \nabla h_j(\varphi(t)) - \mu(t)\varphi(t) \equiv 0$; and
(e) $\nu_j(t) h_j \circ \varphi(t) \equiv 0, j = 1, \ldots, m$.

Taking the derivative in t of the function $f \circ \varphi(t)$ and using conditions (a)–(e), we deduce that

$$0 = \frac{d}{dt}(f \circ \varphi)(t) = \frac{\mu(t)}{2} \frac{d\|\varphi(t)\|^2}{dt}.$$

So $\mu(t) \equiv 0$, and hence $c \in K_0(f, S)$. □

The next theorem says that the study of the problem of minimizing f on S can be reduced to the study of the problem of minimizing f on $\Gamma(f, S)$.

Theorem 2.4. *We have*

$$\inf\{f(x) \,|\, x \in S\} = \inf\{f(x) \,|\, x \in \Gamma(f, S)\}.$$

Proof. Since $\Gamma(f, S) \subset S$, it suffices to show the inequality:

$$f_* := \inf\{f(x) \,|\, x \in S\} \geq \inf\{f(x) \,|\, x \in \Gamma(f, S)\}.$$

To do this, we first assume that f attains its infimum f_* on S at $x^* \in S$. It follows from the Fritz-John optimality conditions that there exist real numbers κ, λ_i, and ν_j, not all zero, such that

$$\kappa \nabla f(x^*) - \sum_{i=1}^{l} \lambda_i \nabla g_i(x^*) - \sum_{j=1}^{m} \nu_j \nabla h_j(x^*) = 0,$$

$$\nu_j h_j(x^*) = 0, \quad j = 1, \ldots, m.$$

Hence, $x^* \in \Gamma(f, S)$ and there is nothing to prove.

We now assume that f does not attain its infimum f_* on S. Then there is a sequence $\{x^k\}_{k \in \mathbb{N}} \subset S$ such that $\lim_{k \to \infty} f(x^k) = f_*$. For each $k \in \mathbb{N}$, since $S_k := \{x \in S \,|\, \|x\|^2 = \|x^k\|^2\}$ is a non-empty compact set, there is $y^k \in S_k$ such that $f(y^k) = \min_{x \in S_k} f(x) \leq f(x^k)$. Hence, $\lim_{k \to \infty} f(y^k) = f_*$. Furthermore, thanks to the Fritz-John optimality conditions, we get $y^k \in \Gamma(f, S)$, and the theorem follows. □

2.3.2 *Tangency values*

Definition 2.4. By the *set of tangency values (at infinity) of f on S*, we mean the set

$$T_\infty(f, S) := \{y \in \mathbb{R} \mid \text{there exists a sequence } x^k \in \Gamma(f, S) \text{ such that}$$
$$\|x^k\| \to +\infty \text{ and } f(x^k) \to y\}.$$

In the case $S = \mathbb{R}^n$, we will write $T_\infty(f)$ instead of $T_\infty(f, S)$.

An immediate corollary of Theorem 2.4 is the following observation.

Corollary 2.1. *Assume that f is bounded from below on S, i.e. $f_* := \inf\{f(x) \mid x \in S\} > -\infty$. If f does not attain its infimum f_* on S, then $f_* \in T_\infty(f, S)$.*

Definition 2.5. We say that f is *coercive* on S if and only if $f(x) \to +\infty$ when $x \in S$ and $\|x\| \to +\infty$.

The following result gives a necessary and sufficient condition for the coercivity of f on S.

Theorem 2.5. *Assume the set S is regular. The following conditions are equivalent:*

(i) *f is coercive on S.*
(ii) *f is bounded from below on S and $T_\infty(f, S) = \emptyset$.*

Proof. (i) $\Rightarrow$ (ii) is immediate.

(ii) $\Rightarrow$ (i) Suppose, by contradiction, that f is not coercive on S. Then there is a sequence $\{x^k\}_{k \in \mathbb{N}} \subset S$ such that $\lim_{k \to \infty} \|x^k\| = +\infty$ and $\lim_{k \to \infty} f(x^k) = c$ for some $c \in \mathbb{R}$. For each $k \in \mathbb{N}$, $S_k := \{x \in S \mid \|x\|^2 = \|x^k\|^2\}$ is a non-empty compact set, and hence there is

$y^k \in S_k$ such that $f(y^k) = \min_{x \in S_k} f(x) \leq f(x^k)$. By the Fritz-John optimality conditions, then $y^k \in \Gamma(f, S)$. Note that the sequence $\{f(y^k)\}_{k \in \mathbb{N}}$ is bounded, so it has an accumulation point, say $c' \in \mathbb{R}$. Clearly, $c' \in T_\infty(f, S)$, which is a contradiction. $\qquad\square$

The next result will play an important role in Chapter 8.

Theorem 2.6. *If the set S is regular, then $T_\infty(f, S)$ is a finite set.*

Proof. Let $\overline{G}$ be the closure of the set

$$G := \{(x, c) \in \mathbb{R}^n \times \mathbb{R} \mid c = f(x), \ x \in \Gamma(f, S)\} \subset \mathbb{R}^n \times \mathbb{R}$$

in $\mathbb{P}^n \times \mathbb{R}$, where $\mathbb{P}^n$ is the real projective space. Then, the sets G and $\overline{G}$ are semi-algebraic. Hence, the set $\overline{G} \setminus G$ is semi-algebraic. Moreover, if $\pi \colon \mathbb{P}^n \times \mathbb{R} \to \mathbb{R}, (u, c) \mapsto c$, is the canonical projection we have $T_\infty(f, S) = \pi(\overline{G} \setminus G)$. Then the Tarski–Seidenberg theorem (Theorem 1.5) implies that $T_\infty(f, S)$ is a semi-algebraic set. Hence, $T_\infty(f, S)$ is a finite union of points and intervals.

By Theorem 2.3, it suffices to show that the set $T_\infty(f, S) \setminus K_0(f, S)$ is finite.

Suppose on the contrary that $T_\infty(f, S) \setminus K_0(f, S)$ contains an open interval I_1. By Lemma 2.5, the semi-algebraic function

$$\eta \colon I_1 \to \mathbb{R}, \qquad r \mapsto \sup\{\|x\|^2 \mid f(x) = r \text{ and } x \in \Gamma(f, S)\},$$

is well defined. Thus, by Theorem 1.8, there exists a non-empty open interval $I_2 \subset I_1$ such that the restriction of η to I_2 is either constant or strictly monotone. Consequently, there is a non-empty open interval $I_3 \subset I_2$ such that the set $\eta(I_3)$ is bounded. But given $c \in I_3 \subset T_\infty(f, S)$, by definition, there exists a continuous semi-algebraic curve $\varphi \colon (0, \epsilon] \to \mathbb{R}^n$, whose image is contained in $\Gamma(f, S)$, with $\|\varphi(t)\| \to +\infty$ and $f \circ \varphi(t) \to c$ as $t \to 0^+$. So there exists $0 < \delta < \epsilon$ such that if $0 < t < \delta$, then $f \circ \varphi(t) \in I_3$. This is impossible since the set

$$\eta(I_3) = \{\|x\|^2 \mid x \in f^{-1}(I_3) \cap \Gamma(f, S)\}$$

is bounded, and the theorem follows. $\qquad\square$

Bibliographic notes

This chapter is inspired mainly by the papers of Hà and Phạm [47, 48, 50–53].

The Fritz-John optimality conditions and the Karush, Kuhn and Tucker optimality conditions can be found in many textbooks of nonlinear programming, for example [134]. Theorem 2.3 also appears in [24, 110]. The notion of a curve of tangencies was introduced in [36]. The tangency variety and the tangency values play an important role in studying of singularities at infinity of polynomials, see [47–49, 92] for more details. Corollary 2.1 was mentioned in [19, Remark 1]. The proof of Theorem 2.6 is adapted from [92]; another proof is given in [47, 48, 51].

Chapter 3

Hölderian Error Bounds
for Polynomial Systems

Abstract

A Hölderian error bound for a subset of an Euclidean space is an inequality that bounds the distance from points in a test set to the given set in terms of a residual function. By exploiting the imposed polynomial structure, we establish a non-smooth extension of the Łojasiewicz gradient inequality to maxima of polynomials as well as bounded Hölderian error bounds for basic closed semi-algebraic sets with exponents explicitly determined by the dimension of the underlying space and the number/degree of the involved polynomials. We also provide some conditions for a semi-algebraic set to have a global Hölderian error bound. We show the existence of global Hölderian error bounds with explicitly calculated exponents for systems of polynomials, which are convenient and non-degenerate at infinity. Finally, a global Hölderian error bound with explicit exponent for one quadratic inequality is given.

3.1 Preliminaries

In this section, we briefly review some definitions and results from variational analysis that will be needed throughout the chapter.

3.1.1 *Subdifferentials*

We begin with the definition of subdifferentials, i.e. an appropriate multivalued operator playing the role of the usual gradient map.

Definition 3.1. Let $f \colon \mathbb{R}^n \to \mathbb{R}$ be a continuous function.

(i) The *Fréchet subdifferential* or *regular subdifferential* of f at $x \in \mathbb{R}^n$, denoted by $\hat{\partial} f(x)$, is given by

$$\hat{\partial} f(x) := \left\{ v \in \mathbb{R}^n \,\middle|\, \liminf_{\|h\| \to 0, \, h \neq 0} \frac{f(x+h) - f(x) - \langle v, h \rangle}{\|h\|} \geq 0 \right\}.$$

(ii) The *limiting subdifferential* or *Mordukhovich subdifferential* of f at $x \in \mathbb{R}^n$, denoted by $\partial f(x)$, is defined to be the set of all cluster points of sequences $\{v^k\}_{k \geq 1}$ such that $v^k \in \hat{\partial} f(x^k)$ and $x^k \to x$ as $k \to \infty$.

Remark 3.1. It is a well-known result of variational analysis that $\hat{\partial} f(x)$ (and a fortiori $\partial f(x)$) is not empty in a dense subset of the domain of f.

Example 3.1. For each $x^0 \in \mathbb{R}^n$, we have

$$\partial(\| \cdot - x^0 \|)(x) = \begin{cases} \dfrac{x - x^0}{\|x - x^0\|} & \text{if } x \neq x^0, \\[2ex] \overline{\mathbb{B}} & \text{otherwise,} \end{cases}$$

where $\overline{\mathbb{B}}$ denotes the closed unit ball centered at the origin in $\mathbb{R}^n$.

Definition 3.2. Using the limiting subdifferential ∂f, we define the *non-smooth slope* of f by

$$\mathfrak{m}_f(x) := \inf\{\|v\| \mid v \in \partial f(x)\}.$$

By definition, $\mathfrak{m}_f(x) = +\infty$ whenever $\partial f(x) = \emptyset$.

Remark 3.2. (i) If f is of class C^1, then $\partial f(x) = \hat{\partial} f(x) = \{\nabla f(x)\}$, and hence $\mathfrak{m}_f(x) = \|\nabla f(x)\|$.

(ii) By the Tarski–Seidenberg theorem (Theorem 1.5), it is not hard to show that if f is semi-algebraic, then so is $\mathfrak{m}_f$.

The limiting subdifferential of continuous functions enjoys the following properties, whose proofs can be found in many textbooks of optimization and variational analysis.

Lemma 3.1. *Let $f\colon \mathbb{R}^n \to \mathbb{R}$ be a continuous function. The following statements are valid:*

(i) *If $x^0 \in \mathbb{R}^n$ is a local (or global) minimum of f then $0 \in \partial f(x^0)$.*

(ii) *Let $x^0 \in \mathbb{R}^n$ be such that $f(x^0) > 0$. Then for any $\rho > 0$,*

$$\partial f^\rho(x^0) = \rho[f(x^0)]^{\rho-1}\partial f(x^0).$$

(iii) *Let $g\colon \mathbb{R}^n \to \mathbb{R}$ be a locally Lipschitz function. Then*

$$\partial(f+g)(x) \subseteq \partial f(x) + \partial g(x) \quad \text{for all } x \in \mathbb{R}^n.$$

Lemma 3.2. *Let $f_1, \ldots, f_p\colon \mathbb{R}^n \to \mathbb{R}$ be functions of class C^1. Then the function $f\colon \mathbb{R}^n \to \mathbb{R}, x \mapsto f(x) := \max_{i=1,\ldots,p} f_i(x)$, is locally Lipschitz and we have for all $x \in \mathbb{R}^n$,*

$$\mathfrak{m}_f(x) = \min_{\lambda_i \geq 0, \ \sum_{i \in I(x)} \lambda_i = 1} \left\| \sum_{i \in I(x)} \lambda_i \nabla f_i(x) \right\|,$$

where $I(x) := \{i \mid f_i(x) = f(x)\}$.

3.1.2 The Ekeland variational principle

To establish error bound results, we recall a classical theorem in analysis which says that if f is a continuous function and bounded from below on $\mathbb{R}^n$, then it possesses a "minimizing sequence". This result, known as the Ekeland variational principle, is valid in general complete metric spaces; for our purpose here, we state the result as a theorem on finite-dimensional Euclidean spaces.

Theorem 3.1 (Ekeland Variational Principle). *Let $f\colon \mathbb{R}^n \to \mathbb{R}$ be a continuous function, bounded from below. Let $\epsilon > 0$ and $x^0 \in \mathbb{R}^n$ be such that*

$$\inf_{x \in \mathbb{R}^n} f(x) \leq f(x^0) \leq \inf_{x \in \mathbb{R}^n} f(x) + \epsilon.$$

Then for any $\lambda > 0$, there exists some point $y^0 \in \mathbb{R}^n$ such that

$$f(y^0) \leq f(x^0),$$

$$\|y^0 - x^0\| \leq \lambda,$$

$$f(y^0) \leq f(x) + \frac{\epsilon}{\lambda}\|x - y^0\| \quad \text{for all } x \in \mathbb{R}^n.$$

Proof. We define the function $g : \mathbb{R}^n \to \mathbb{R}, x \mapsto g(x)$, by

$$g(x) := f(x) + \frac{\epsilon}{\lambda} \|x - x^0\|.$$

Clearly, the function g is continuous and verifies: $\lim_{\|x\| \to +\infty} g(x) = +\infty$. Therefore, there exists y^0 minimizing g on $\mathbb{R}^n$ such that

$$f(y^0) + \frac{\epsilon}{\lambda} \|y^0 - x^0\| \ \leq \ f(x) + \frac{\epsilon}{\lambda} \|x - x^0\| \quad \text{for all } x \in \mathbb{R}^n.$$

By letting $x = x^0$, we get

$$f(y^0) + \frac{\epsilon}{\lambda} \|y^0 - x^0\| \ \leq \ f(x^0).$$

This implies that $f(y^0) \leq f(x^0)$. Now, since $f(x^0) \leq \inf_{x \in \mathbb{R}^n} f(x) + \epsilon \leq f(y^0) + \epsilon$, we clearly deduce from the above that $\|y^0 - x^0\| \leq \lambda$. Moreover, we have for all $x \in \mathbb{R}^n$,

$$f(y^0) \leq f(x) + \frac{\epsilon}{\lambda} \left(\|x - x^0\| - \|y^0 - x^0\| \right)$$

$$\leq f(x) + \frac{\epsilon}{\lambda} \|x - y^0\|,$$

which is the desired third inequality. $\qquad\square$

3.2 Non-smooth Łojasiewicz's gradient inequality with explicit exponent for maxima of finitely many polynomials

The result given below establishes a non-smooth version of Łojasiewicz's gradient inequality for maximum functions over finitely many polynomials. It is certainly of its own interest while being applied in what follows to deriving error bounds with explicit exponents for polynomial systems.

Theorem 3.2. *Let* $f(x) := \max\{f_1(x), \dots, f_p(x)\}$, *where* f_i, $i = 1, \dots, p$, *are polynomials on* $\mathbb{R}^n$ *of degree at most* d, *and let* $\bar{x} \in \mathbb{R}^n$ *with* $f(\bar{x}) = 0$. *Then there are numbers* $c > 0$ *and* $\epsilon > 0$ *such that*

$$\mathfrak{m}_f(x) \geq c |f(x)|^{1 - \frac{1}{\mathscr{R}(n+p-1,\, d+1)}} \quad \text{for all } x \text{ with } \|x - \bar{x}\| \leq \epsilon.$$

Proof. Without loss of generality, assume that $f_i(\bar{x}) = 0$ for all $i = 1, \ldots, p$. Then for each non-empty subset $I := \{i_1, \ldots, i_q\} \subset \{1, \ldots, p\}$, we define the function $F_I : \mathbb{R}^n \times \mathbb{R}^{q-1} \rightarrow \mathbb{R}, (x, \lambda) \mapsto F_I(x, \lambda)$, by

$$F_I(x, \lambda) := \begin{cases} \displaystyle\sum_{j=1}^{q-1} \lambda_j f_{i_j}(x) + \left(1 - \sum_{j=1}^{q-1} \lambda_j\right) f_{i_q}(x) & \text{if } q \geq 2, \\ f_{i_1}(x) & \text{if } q = 1, \end{cases}$$

which is clearly a polynomial on $\mathbb{R}^{n+q-1}$ with degree at most $d+1$ and $F_I(\bar{x}, \lambda) = 0$ for all $\lambda := (\lambda_1, \ldots, \lambda_{q-1}) \in \mathbb{R}^{q-1}$. Define further the set $\mathbf{P} \subset \mathbb{R}^{q-1}$ by

$$\mathbf{P} := \left\{\lambda \in \mathbb{R}^{q-1} \,\middle|\, \lambda_j \geq 0, \ \sum_{j=1}^{q-1} \lambda_j \leq 1\right\}.$$

Since the set $\mathbf{P}$ is compact, it follows from Theorem 1.16 that there exist numbers $c_I > 0$ and $\epsilon_I > 0$ for which we have

$$\|\nabla F_I(x, \lambda)\| \geq c_I \, |F_I(x, \lambda)|^{1 - \frac{1}{\mathscr{R}(n+q-1, d+1)}} \quad \text{whenever}$$

$$\|x - \bar{x}\| \leq \epsilon_I \quad \text{and} \quad \lambda \in \mathbf{P}.$$

Let

$$c := \min\left\{c_I \,\middle|\, \emptyset \neq I \subset \{1, \ldots, p\}\right\} > 0,$$

$$\epsilon := \min\left\{\epsilon_I \,\middle|\, \emptyset \neq I \subset \{1, \ldots, p\}\right\} > 0.$$

Pick an arbitrary point x in $\mathbb{R}^n$ with $\|x - \bar{x}\| \leq \epsilon$ and denote $\bar{I} = I(x) := \{i \,|\, f_i(x) = f(x)\}$. Lemma 3.2 tells us that there are numbers $\lambda_i \geq 0$ for $i \in \bar{I}$ such that $\sum_{i \in \bar{I}} \lambda_i = 1$ and

$$\mathfrak{m}_f(x) = \left\|\sum_{i \in \bar{I}} \lambda_i \nabla f_i(x)\right\|.$$

By renumbering, if necessary, we may assume that $\bar{I} = \{i_1, \ldots, i_{\bar{q}}\}$, where $\bar{q}$ signifies its cardinality. We only consider the case where

$\bar{q} \geq 2$; the case $\bar{q} = 1$ is proved similarly. By definition, we have

$$F_{\bar{I}}(x, \lambda_{i_1}, \ldots, \lambda_{i_{\bar{q}-1}})$$

$$= \sum_{j=1}^{\bar{q}} \lambda_{i_j} f_{i_j}(x) = \sum_{i \in \bar{I}} \lambda_i f_i(x) = \sum_{i \in I(x)} \lambda_i f_i(x) = f(x).$$

Furthermore, it is easy to see that

$$\|\nabla F_{\bar{I}}(x, \lambda_{i_1}, \ldots, \lambda_{i_{\bar{q}-1}})\|$$

$$= \left\| \left(\sum_{j=1}^{\bar{q}} \lambda_{i_j} \nabla f_{i_j}(x), f_{i_1}(x) - f_{i_{\bar{q}}}(x), \ldots, f_{i_{\bar{q}-1}}(x) - f_{i_{\bar{q}}}(x) \right) \right\|$$

$$= \left\| \sum_{j=1}^{\bar{q}} \lambda_{i_j} \nabla f_{i_j}(x) \right\| = \left\| \sum_{i \in I(x)} \lambda_i \nabla f_i(x) \right\| = \mathfrak{m}_f(x).$$

Therefore, we deduce successively

$$\mathfrak{m}_f(x) = \|\nabla F_{\bar{I}}(x, \lambda_{i_1}, \ldots, \lambda_{i_{\bar{q}-1}})\|$$

$$\geq c_{\bar{I}} |F_{\bar{I}}(x, \lambda_{i_1}, \ldots, \lambda_{i_{\bar{q}-1}})|^{1 - \frac{1}{\mathscr{R}(n+\bar{q}-1, d+1)}}$$

$$= c_{\bar{I}} |f(x)|^{1 - \frac{1}{\mathscr{R}(n+\bar{q}-1, d+1)}}$$

$$\geq c |f(x)|^{1 - \frac{1}{\mathscr{R}(n+p-1, d+1)}},$$

which completes the proof of the theorem. □

3.3 Bounded Hölderian error bounds with explicit exponents for basic closed semi-algebraic sets

Let g_i as $i = 1, \ldots, l$ and h_j as $j = 1, \ldots, m$, be continuous functions on $\mathbb{R}^n$, and let S denote the solution set of the system of equalities and inequalities:

$$g_1(x) = 0, \ldots, g_l(x) = 0, \quad h_1(x) \leq 0, \ldots, h_m(x) \leq 0.$$

For any $x \in \mathbb{R}^n$, if $x \notin S$, then some equality or inequality constraints in this system must be violated at x. The amount of "constraint violation" at x, called its *residual*, is given by $\sum_{i=1}^{l} |g_i(x)| + \sum_{j=1}^{m} [h_j(x)]_+$, where we set $[h_j(x)]_+ := \max\{h_j(x), 0\}$. We are interested in obtaining inequalities that bound the distance $\text{dist}(x, S)$ from x in a given set K to S in terms of the residual at x.

Definition 3.3. We say that a *bounded Hölderian error bound* holds for the set S if for any compact set $K \subset \mathbb{R}^n$, there exist some positive constants c and α such that

$$c \, \text{dist}(x, S) \leq \left(\sum_{i=1}^{l} |g_i(x)| + \sum_{j=1}^{m} [h_j(x)]_+ \right)^\alpha \quad \text{for all } x \in K.$$

We remark that if all the functions g_i and h_j (defining S) are semi-algebraic, then a bounded Hölderian error bound holds for S; this is an immediate consequence of Corollary 1.2. Furthermore, Theorem 3.3, stated below, shows that the exponent α can be determined explicitly when all g_i and h_j are polynomials. To proceed, we need the following lemma.

Lemma 3.3. *Let $f \colon \mathbb{R}^n \to \mathbb{R}$ be a continuous function and let $\bar{x} \in \mathbb{R}^n$ be such that $f(\bar{x}) = 0$. Let $S := \{x \in \mathbb{R}^n \mid f(x) \leq 0\}$. Assume that there are real numbers $c > 0, \delta > 0$, and $\rho \in [0, 1)$, such that*

$$\mathfrak{m}_f(x) \geq c|f(x)|^\rho \quad \text{for all } \|x - \bar{x}\| \leq \delta \text{ and } x \notin S.$$

Then we have

$$[f(x)]_+^{1-\rho} \geq c(1 - \rho) \, \text{dist}(x, S) \quad \text{whenever } \|x - \bar{x}\| \leq \frac{\delta}{2},$$

where $[f(x)]_+ := \max\{f(x), 0\}$ and $\text{dist}(x, S)$ denotes the Euclidean distance from x to S.

Proof. By contradiction, suppose that there is $x^0 \in \mathbb{R}^n$, with $\|x^0 - \bar{x}\| \leq \frac{\delta}{2}$, such that

$$[f(x^0)]_+^{1-\rho} < c(1 - \rho) \, \text{dist}(x^0, S).$$

Then $x^0 \notin S$ and $f(x^0) > 0$. Moreover there exists $c_0 \in (0, 1)$ such that

$$[f(x^0)]_+^{1-\rho} < c_0 c(1 - \rho) \, \text{dist}(x^0, S) < c(1 - \rho) \, \text{dist}(x^0, S).$$

Note that $\inf_{x \in \mathbb{R}^n} [f(x)]_+^{1-\rho} = 0$. Let

$$\epsilon := [f(x^0)]_+^{1-\rho} > 0 \quad \text{and} \quad \lambda := c_0 \operatorname{dist}(x^0, S) > 0.$$

By the Ekeland Variational Principle (Theorem 3.1), there exists $y^0 \in \mathbb{R}^n$ such that the following inequalities hold

$$[f(y^0)]_+^{1-\rho} \le [f(x^0)]_+^{1-\rho},$$

$$\|y^0 - x^0\| \le \lambda,$$

$$[f(y^0)]_+^{1-\rho} \le [f(x)]_+^{1-\rho} + \frac{\epsilon}{\lambda}\|x - y^0\| \quad \text{for all } x \in \mathbb{R}^n.$$

Consequently, $\|y^0 - x^0\| \le \lambda < \operatorname{dist}(x^0, S)$, and so $y^0 \notin S$ and $f(y^0) > 0$. Moreover, we have

$$\|y^0 - \bar{x}\| \le \|y^0 - x^0\| + \|x^0 - \bar{x}\| \le \lambda + \frac{\delta}{2} < \operatorname{dist}(x^0, S) + \frac{\delta}{2}$$

$$\le \|x^0 - \bar{x}\| + \frac{\delta}{2} \le \delta.$$

By continuity, f is positive in some open neighborhood of y^0. Therefore, for all x near y^0, we have $[f(x)]_+ = f(x)$ and y^0 is a local minimizer of the function

$$\mathbb{R}^n \to \mathbb{R}, \quad x \mapsto [f(x)]_+^{1-\rho} + \frac{\epsilon}{\lambda}\|x - y^0\|.$$

From Lemma 3.1, we deduce successively

$$0 \in \partial\left([f(\cdot)]_+^{1-\rho} + \frac{\epsilon}{\lambda}\| \cdot - y^0\|\right)(y^0) \subseteq \partial([f(\cdot)]_+^{1-\rho})(y^0) + \frac{\epsilon}{\lambda}\overline{\mathbb{B}}$$

$$= (1-\rho)[f(y^0)]^{-\rho}\partial f(y^0) + \frac{\epsilon}{\lambda}\overline{\mathbb{B}}.$$

By definition, then

$$(1-\rho)[f(y^0)]^{-\rho}\,\mathfrak{m}_f(y^0) \le \frac{\epsilon}{\lambda} = \frac{[f(x^0)]_+^{1-\rho}}{c_0 \operatorname{dist}(x^0, S)}$$

$$< \frac{c_0 c(1-\rho)\operatorname{dist}(x^0, S)}{c_0 \operatorname{dist}(x^0, S)} = c(1-\rho),$$

which is a contradiction. $\qquad\square$

Prior to deriving the main results of this section, we present an example illustrating the dependence of error bounds for polynomial systems on the degree of the polynomials involved and on the dimension of the problem/space in question.

Example 3.2. Let $d \in \mathbb{N}$, and let $g_i(x_1, \ldots, x_n) := x_{i+1} - x_i^d$ for $i = 1, \ldots, n-1$, and $g_n(x_1, \ldots, x_n) := x_n^d$. We denote by S the solution set of the system

$$g_1(x) = 0, \ldots, g_n(x) = 0.$$

Clearly, $0 \in \mathbb{R}^n$ is the unique solution of this system, so $S = \{0\}$. Consider the curve $x(\tau) := (\tau, \tau^d, \ldots, \tau^{d^{n-1}}) \in \mathbb{R}^n$, where $\tau > 0$ is a small parameter. It is easy to see that

$$\mathrm{dist}(x(\tau), S) = \sqrt{\sum_{i=1}^{n} \tau^{2d^{i-1}}} = O(\tau) \quad \text{and} \quad \sum_{i=1}^{n} |g_i(x(\tau))| = \tau^{d^n}.$$

Thus, we have

$$\mathrm{dist}\big(x(\tau), S\big) = O\left(\left[\sum_{i=1}^{n} |g_i(x(\tau))|\right]^{\frac{1}{d^n}}\right),$$

which shows that the exponent α in any bounded Hölderian error bound for S does not exceed d^{-n}.

Now, we are ready to derive bounded Hölderian error bounds with explicit exponents for basic closed semi-algebraic sets.

Theorem 3.3. *Let g_i, as $i = 1, \ldots, l$, and h_j, as $j = 1, \ldots, m$, be real polynomials on $\mathbb{R}^n$ of degree at most d, and let*

$$S := \{x \in \mathbb{R}^n \mid g_1(x) = 0, \ldots, g_l(x) = 0,$$

$$h_1(x) \leq 0, \ldots, h_m(x) \leq 0\} \neq \emptyset.$$

Then for any compact set $K \subset \mathbb{R}^n$, there is a constant $c > 0$ such that

$$c \, \mathrm{dist}(x, S)$$

$$\leq \left(\sum_{i=1}^{l} |g_i(x)| + \sum_{j=1}^{m} [h_j(x)]_+\right)^{\frac{1}{\mathscr{R}(n+l+m-1, d+1)}} \qquad \textit{for all } x \in K.$$

Proof. Since K is compact, it suffices to show that for each $\bar{x} \in \mathbb{R}^n$, there are real numbers $c > 0$ and $\epsilon > 0$ such that for all $\|x - \bar{x}\| \leq \epsilon$,

$$c \operatorname{dist}(x, S) \leq \left(\sum_{i=1}^{l} |g_i(x)| + \sum_{j=1}^{m} [h_j(x)]_+ \right)^{\frac{1}{\mathscr{R}(n+l+m-1, d+1)}}.$$

To do this, define the function $f \colon \mathbb{R}^n \to \mathbb{R}$ by

$$f(x) := \max \left\{ |g_1(x)|, \ldots, |g_l(x)|, [h_1(x)]_+, \ldots, [h_m(x)]_+ \right\}.$$

Clearly, f is a non-negative and continuous function on $\mathbb{R}^n$ and $S = \{x \in \mathbb{R}^n \mid f(x) = 0\}$. The desired conclusion is rather straightforward if $f(\bar{x}) > 0$. Hence, we assume that $f(\bar{x}) = 0$. Then $g_i(\bar{x}) = 0$ for $i = 1, \ldots, l$, and $h_j(\bar{x}) \leq 0$ for $j = 1, \ldots, m$.

By definition, we have

$$f(x) = \max \big\{ g_1(x), \ldots, g_l(x), -g_1(x), \ldots, -g_l(x),$$
$$h_1(x), \ldots, h_m(x) \big\}.$$

For each $e := (e_1, \ldots, e_l) \in \{-1, 1\}^l$, we define the function $f_e \colon \mathbb{R}^n \to \mathbb{R}$ by

$$f_e(x) := \max \big\{ e_1 g_1(x), \ldots, e_l g_l(x), h_1(x), \ldots, h_m(x) \big\},$$

which is the maximum of $l+m$ polynomials of degree at most d. Note that $f_e(\bar{x}) = 0$. Theorem 3.2 gives us numbers $c(e) > 0$ and $\epsilon(e) > 0$ such that

$$\mathfrak{m}_{f_e}(x) \geq c(e) |f_e(x)|^{1 - \frac{1}{\mathscr{R}(n+l+m-1, d+1)}} \quad \text{whenever } \|x - \bar{x}\| \leq \epsilon(e).$$

Let

$$c' := \min_{e \in \{-1, 1\}^m} c(e) > 0 \quad \text{and} \quad \epsilon := \min_{e \in \{-1, 1\}^m} \epsilon(e) > 0.$$

Take any x with $\|x - \bar{x}\| \leq \epsilon$ and $f(x) > 0$. Then for each $i = 1, \ldots, l$, we have that either $g_i(x) \neq -g_i(x)$ or $g_i(x) < f(x)$. It allows us to find $e \in \{-1, 1\}^l$ so that $f(x) = f_e(x)$ and $\mathfrak{m}_f(x) = \mathfrak{m}_{f_e}(x)$. This gives us the following estimate:

$$\mathfrak{m}_f(x) = \mathfrak{m}_{f_e}(x) \geq c(e) \, |f_e(x)|^{1 - \frac{1}{\mathscr{R}(n+l+m-1,d+1)}}$$

$$\geq c' \, |f(x)|^{1 - \frac{1}{\mathscr{R}(n+l+m-1,d+1)}},$$

which completes the proof of the theorem by applying Lemma 3.3.

$\square$

3.4 Criteria for the existence of global Hölderian error bounds

Let $f \colon \mathbb{R}^n \to \mathbb{R}$ be a continuous semi-algebraic function. Assume that $S := \{x \in \mathbb{R}^n \, | \, f(x) \leq 0\} \neq \emptyset$. Let $[f(x)]_+ := \max\{f(x), 0\}$. Then $x \in S$ if and only if $[f(x)]_+ = 0$. By Corollary 1.2, for each compact subset K in $\mathbb{R}^n$, there exist constants $c > 0$ and $\alpha > 0$ such that the following inequality holds:

$$c \, \mathrm{dist}(x, S) \leq [f(x)]_+^\alpha \quad \text{for all } x \in K. \tag{3.1}$$

Thus, (3.1) provides a simple upper estimate for how far a point is away from the set S. The following examples indicate that in general (3.1) cannot hold globally for all $x \in \mathbb{R}^n$.

Example 3.3. (i) Let $f(x, y) := x(y^2 + (xy - 1)^2)$. Then $S := \{(x, y) \in \mathbb{R}^2 \, | \, f(x, y) \leq 0\} = \{x \leq 0\}$. Consider the sequence $z^k := (k, \frac{1}{k}) \in \mathbb{R}^2$ for $k \geq 1$. We have $f(z^k) = \frac{1}{k} \to 0$ and $\mathrm{dist}(z^k, S) = k \to +\infty$ as $k \to +\infty$. Thus, (3.1) cannot hold for all $(x, y) \in \mathbb{R}^2$.

(ii) Let $f(x, y) := (y - 1)^2 + (xy - 1)^2$. Then $S := \{(x, y) \in \mathbb{R}^2 \, | \, f(x, y) \leq 0\} = \{(1, 1)\}$. Consider the sequence $z^k := (k, \frac{1}{k})$ for $k \geq 1$. It is clear that $f(z^k) = (\frac{1}{k} - 1)^2 \to 1$ and $\mathrm{dist}(z^k, S) = (k - 1)^2 + (\frac{1}{k} - 1)^2 \to +\infty$ as $k \to +\infty$. Consequently, (3.1) cannot hold for all $(x, y) \in \mathbb{R}^2$.

Notably, in the above examples, the following implications fail:
$$[f(x)]_+ \to 0 \Longrightarrow \mathrm{dist}(x, S) \to 0,$$
$$\mathrm{dist}(x, S) \to +\infty \Longrightarrow [f(x)]_+ \to +\infty.$$
These observations lead to the following two lemmas.

Lemma 3.4 (Hölderian error bound "near to S"). *The following two statements are equivalent:*

(i) *For any sequence $x^k \in \mathbb{R}^n \setminus S$, with $\|x^k\| \to +\infty$, it holds that*
$$f(x^k) \to 0 \Longrightarrow \mathrm{dist}(x^k, S) \to 0.$$

(ii) *There exist some constants $c > 0, \delta > 0$, and $\alpha > 0$ such that*
$$c\,\mathrm{dist}(x, S) \le [f(x)]_+^\alpha \quad \text{for all } x \in f^{-1}((-\infty, \delta]).$$

Proof. (ii) $\Rightarrow$ (i) The implication is clear.

(i) $\Rightarrow$ (ii) The lemma is trivial in the case $S = \mathbb{R}^n$, so assume that $S \ne \mathbb{R}^n$. Then, by the continuity of f, there exists $\delta_1 > 0$ such that $f^{-1}(t) \ne \emptyset$ for all $t \in [0, \delta_1]$. Let
$$\mu(t) := \sup_{x \in f^{-1}(t)} \mathrm{dist}(x, S) \quad \text{for } t \in [0, \delta_1].$$
The condition (i) implies that there exists $0 < \delta_2 \le \delta_1$ such that $\mu(t) < +\infty$ for all $t \in [0, \delta_2]$ and $\mu(t) \to 0$ when $t \to 0^+$. In view of the Tarski–Seidenberg theorem (Theorem 1.5), the function $[0, \delta_2] \to \mathbb{R}, t \mapsto \mu(t)$, is semi-algebraic. By the Growth Dichotomy Lemma (Lemma 1.7), we can write
$$\mu(t) = at^\alpha + \text{higher order terms in } t, \quad 0 \le t \ll 1,$$
for some constants $a > 0$ and $\alpha > 0$. We deduce that there exist $c > 0$ and $0 < \delta \le \delta_2$ such that
$$c\,\mathrm{dist}(x, S) \le [f(x)]_+^\alpha \quad \text{for all } x \in f^{-1}((-\infty, \delta]),$$
which completes the proof of the lemma. $\qquad\qquad\square$

Lemma 3.5 (Hölderian error bound "far from S"). *Suppose that for any sequence $x^k \in \mathbb{R}^n \setminus S$, with $\|x^k\| \to +\infty$, it holds that*
$$\mathrm{dist}(x^k, S) \to +\infty \Longrightarrow f(x^k) \to +\infty.$$
Then there exist some constants $c > 0, r > 0$, and $\beta > 0$ such that
$$c\,\mathrm{dist}(x, S) \le [f(x)]_+^\beta \quad \text{for all } x \in f^{-1}([r, +\infty)).$$

Proof. The lemma is trivial in the case $S = \mathbb{R}^n$, so assume that $S \neq \mathbb{R}^n$.

We first assume that f is bounded from above, say by $r > 0$. It follows from the assumption that there exists $M > 0$ such that $\operatorname{dist}(x, S) \leq M$ for all $x \in \mathbb{R}^n$. Then we have for all $x \in f^{-1}([r, +\infty))$,

$$f(x) \geq r = \frac{r}{M}M \geq \frac{r}{M}\operatorname{dist}(x, S),$$

which implies the lemma.

We now suppose that f is not bounded from above. Since $S \neq \emptyset$ and by the continuity of f, we get $f^{-1}(t) \neq \emptyset$ for any $t \geq 0$. Let

$$\mu(t) := \sup_{x \in f^{-1}(t)} \operatorname{dist}(x, S), \quad \text{for } t \geq 0.$$

The assumption implies that there exists $r_1 \geq 0$ such that $\mu(t) < +\infty$ for all $t \in [r_1, +\infty)$. Thanks to the Tarski–Seidenberg theorem (Theorem 1.5), the function $[r_1, +\infty) \to \mathbb{R}, t \mapsto \mu(t)$, is semi-algebraic. By the Growth Dichotomy Lemma (Lemma 1.7), there exist scalars $b > 0$ and $\beta \in \mathbb{Q}$ such that

$$\mu(t) = bt^\beta + \text{lower order terms in } t, \quad t \gg 1,$$

Let

$$M := \sup_{t \in [r_1, +\infty)} \mu(t).$$

We distinguish two cases.

Case 1: $M = +\infty$.

Then $\lim_{t \to +\infty} \mu(t) = +\infty$. Therefore $\beta > 0$, and so we find constants $c > 0$ and $r > r_1$ such that

$$c\operatorname{dist}(x, S) \leq [f(x)]_+^\beta \quad \text{for all } x \in f^{-1}([r, +\infty)),$$

which proves the lemma.

Case 2: $M < +\infty$.

Take any $r \geq r_1$. We have for all $x \in f^{-1}([r, +\infty))$,

$$f(x) \geq r = \frac{r}{M}M \geq \frac{r}{M}\operatorname{dist}(x, S).$$

The lemma now follows easily. $\qquad\square$

Definition 3.4. We say that the function f (or the set $S := \{x \in \mathbb{R}^n \mid f(x) \le 0\}$) admits a *global Hölderian error bound* if there exist positive scalars c, α and β such that

$$c \operatorname{dist}(x, S) \le [f(x)]_+^\alpha + [f(x)]_+^\beta \quad \text{for all } x \in \mathbb{R}^n.$$

The following result gives a necessary and sufficient condition for a continuous semi-algebraic to have a global Hölderian error bound.

Theorem 3.4. *The following two statements are equivalent:*

(i) *For any sequence $x^k \in \mathbb{R}^n \setminus S$, with $\|x^k\| \to +\infty$, we have*

    (i1) *if $f(x^k) \to 0$ then $\operatorname{dist}(x^k, S) \to 0$; and*
    (i2) *if $\operatorname{dist}(x^k, S) \to +\infty$ then $f(x^k) \to +\infty$.*

(ii) *The function f has a global Hölderian error bound.*

Proof. (ii) $\Rightarrow$ (i) The implication is straightforward.

(i) $\Rightarrow$ (ii) By Lemmas 3.4 and 3.5, there exist positive constants c_1, c_2, δ, and r, and exponents $\alpha > 0, \beta > 0$ such that

$$c_1 \operatorname{dist}(x, S) \le [f(x)]_+^\alpha \quad \text{for all } x \in f^{-1}((-\infty, \delta])$$

and

$$c_2 \operatorname{dist}(x, S) \le [f(x)]_+^\beta \quad \text{for all } x \in f^{-1}([r, +\infty)).$$

On the other hand, it follows easily from the condition (i2) that there is a constant $M > 0$ such that for all $x \in f^{-1}([\delta, r])$, we have $\operatorname{dist}(x, S) \le M$ and hence

$$[f(x)]_+^\alpha + [f(x)]_+^\beta \ge \delta^\alpha + \delta^\beta = \frac{\delta^\alpha + \delta^\beta}{M} M \ge \frac{\delta^\alpha + \delta^\beta}{M} \operatorname{dist}(x, S).$$

Letting $c := \min\{c_1, c_2, \frac{\delta^\alpha + \delta^\beta}{M}\}$, we get the desired conclusion. $\square$

The following simple example shows that the exponents α and β in global Hölderian error bounds are different in general.

Example 3.4. Let $n := 2$ and $f(x, y) := x^2 + y^4$. Then $S := \{(x, y) \in \mathbb{R}^2 \mid f(x, y) \le 0\} = \{(0, 0)\}$ and so $\operatorname{dist}((x, y), S) = \|(x, y)\|$. Furthermore, it is not hard to see that there are no constants $c > 0$ and

$\alpha > 0$ such that

$$c\,\text{dist}((x,y), S) \le [f(x,y)]_+^\alpha \quad \text{for all } (x,y) \in \mathbb{R}^2.$$

On the other hand, it holds that

$$2\,\text{dist}((x,y), S) \le [f(x,y)]_+^{\frac{1}{2}} + [f(x,y)]_+^{\frac{1}{4}} \quad \text{for all } (x,y) \in \mathbb{R}^2.$$

3.5 The Palais–Smale condition and global Hölderian error bounds

In this section, we describe a relation between the Palais–Smale condition and the existence of global Hölderian error bounds for continuous semi-algebraic functions. Let us start with the following definition.

Definition 3.5. Given a continuous function $f \colon \mathbb{R}^n \to \mathbb{R}$ and a real number t, we say that f satisfies the *Palais–Smale condition at the level t*, if every sequence $\{x^k\}_{k \in \mathbb{N}} \subset \mathbb{R}^n$ such that $f(x^k) \to t$ and $\mathfrak{m}_f(x^k) \to 0$ as $k \to \infty$ possesses a convergence subsequence.

The fact that the Palais–Smale condition implies the existence of an error bound is well known as well as its proof technique via the Ekeland variational principle. This kind of results can be established in very general settings, for continuous functions on metric spaces. Here, using semi-algebraic properties of considered functions, we obtain an error bound in a form, which is more convenient for our further investigation.

Theorem 3.5. *Let $f \colon \mathbb{R}^n \to \mathbb{R}$ be a continuous semi-algebraic function. Assume that $S := \{x \in \mathbb{R}^n \mid f(x) \le 0\} \ne \emptyset$. If f satisfies the Palais–Smale condition at each level $t \ge 0$, then f admits a global Hölderian error bound, i.e. there exist some constants $c > 0, \alpha > 0$, and $\beta > 0$ such that*

$$c\,\text{dist}(x, S) \le [f(x)]_+^\alpha + [f(x)]_+^\beta \quad \text{for all } x \in \mathbb{R}^n.$$

Proof. It is sufficient to show that the condition (i) in Theorem 3.4 holds. To see this, we proceed by the method of contradiction.

We first assume that there exist a number $\delta > 0$ and a sequence $x^k \in \mathbb{R}^n \setminus S$, with $\|x^k\| \to +\infty$, such that

$$f(x^k) \to 0 \quad \text{and} \quad \text{dist}(x^k, S) \geq \delta.$$

Since $S \neq \emptyset$ and $[f(x)]_+ \geq 0$ for $x \in \mathbb{R}^n$, we have $\inf_{x \in \mathbb{R}^n}[f(x)]_+ = 0$. Applying the Ekeland Variational Principle (Theorem 3.1) to the function $\mathbb{R}^n \to \mathbb{R}, x \mapsto [f(x)]_+$, with data $\epsilon := [f(x^k)]_+ = f(x^k) > 0$ and $\lambda := \frac{\delta}{2} > 0$, we find a point y^k in $\mathbb{R}^n$ such that the following inequalities hold:

$$[f(y^k)]_+ \leq [f(x^k)]_+,$$
$$\|y^k - x^k\| \leq \lambda,$$
$$[f(y^k)]_+ \leq [f(x)]_+ + \frac{\epsilon}{\lambda}\|x - y^k\| \quad \text{for all } x \in \mathbb{R}^n.$$

We deduce easily that $\lim_{k \to \infty} \|y^k\| = +\infty$ and

$$\text{dist}(y^k, S) \geq \text{dist}(x^k, S) - \|y^k - x^k\|$$
$$\geq \text{dist}(x^k, S) - \lambda = \text{dist}(x^k, S) - \frac{\delta}{2} \geq \frac{\delta}{2}.$$

Hence, $y^k \notin S$ and $f(y^k) > 0$. By continuity, f is positive in some open neighborhood of y^k. In particular, we have for all x near y^k, $[f(x)]_+ = f(x)$ and so y^k is a local minimizer of the function

$$\mathbb{R}^n \mapsto \mathbb{R}, \quad x \mapsto f(x) + \frac{\epsilon}{\lambda}\|x - y^k\|.$$

From Lemma 3.1, we deduce successively

$$0 \in \partial\left(f(\cdot) + \frac{\epsilon}{\lambda}(\| \cdot - y^k\|)\right)(y^k)$$
$$\subset \partial f(y^k) + \frac{\epsilon}{\lambda}\partial(\| \cdot - y^k\|)(y^k) = \partial f(y^k) + \frac{\epsilon}{\lambda}\overline{\mathbb{B}}.$$

Therefore,

$$\mathfrak{m}_f(y^k) \leq \frac{\epsilon}{\lambda} = \frac{2f(x^k)}{\delta}.$$

By letting k tend to infinity, we obtain

$$\lim_{k\to\infty} \|y^k\| = +\infty, \quad \lim_{k\to\infty} f(y^k) = 0, \quad \text{and} \quad \lim_{k\to\infty} \mathfrak{m}_f(y^k) = 0.$$

So, f does not satisfy the Palais–Smale condition at the value $t = 0$, and a contradiction follows.

We next suppose that there exist a number $M > 0$ and a sequence $x^k \in \mathbb{R}^n \setminus S$, with $\|x^k\| \to +\infty$, such that

$$\text{dist}(x^k, S) \to +\infty \quad \text{and} \quad 0 < f(x^k) \le M.$$

Again, we see that $\inf_{x\in\mathbb{R}^n}[f(x)]_+ = 0$. We apply the Ekeland Variational Principle (Theorem 3.1) to the function $\mathbb{R}^n \to \mathbb{R}, x \mapsto [f(x)]_+$, with data $\epsilon := [f(x^k)]_+ = f(x^k) > 0$ and $\lambda := \frac{\text{dist}(x^k,S)}{2} > 0$; there exists a point y^k in $\mathbb{R}^n$ satisfying the following inequalities

$$[f(y^k)]_+ \le [f(x^k)]_+,$$

$$\|y^k - x^k\| \le \lambda,$$

$$[f(y^k)]_+ \le [f(x)]_+ + \frac{\epsilon}{\lambda}\|x - y^k\| \quad \text{for all } x \in \mathbb{R}^n.$$

We deduce easily that

$$\text{dist}(y^k, S) \ge \text{dist}(x^k, S) - \|y^k - x^k\|$$

$$\ge \text{dist}(x^k, S) - \lambda = \frac{\text{dist}(x^k, S)}{2},$$

which yields $\lim_{k\to\infty} \text{dist}(y^k, S) = +\infty$. In particular, we get $y^k \in \mathbb{R}^n \setminus S$ and $\|y^k\| \to +\infty$.

By an argument as above, we can see that

$$\mathfrak{m}_f(y^k) \le \frac{\epsilon}{\lambda} = \frac{2f(x^k)}{\text{dist}(x^k, S)} \le \frac{2M}{\text{dist}(x^k, S)}.$$

Hence,

$$\lim_{k\to\infty} \mathfrak{m}_f(y^k) = 0.$$

Note that $0 < f(y^k) \leq f(x^k) \leq M$ for all $k \geq 1$. Hence, by passing to subsequences if necessary, we may assume that there exists the limit $t := \lim_{k \to \infty} f(y^k) \geq 0$. Therefore, f does not satisfy the Palais–Smale condition at t, which is a contradiction. The proof of Theorem 3.5 is complete. $\qquad\qquad\qquad\qquad\qquad\qquad\qquad\square$

Example 3.5 (0/1 integer feasibility problem). Let g_i as $i = 1, \ldots, l$, and h_j as $j = 1, \ldots, m$, be continuous semi-algebraic functions on $\mathbb{R}^n$. Consider the following 0/1 integer feasibility problem:

$$g_1(x) = 0, \ldots, g_l(x) = 0, \quad h_1(x) \leq 0, \ldots, h_m(x) \leq 0,$$

$$x_i = 0 \quad \text{or} \quad 1, \ i = 1, \ldots, n.$$

Equivalently, we may consider the following system:

$$g_1(x) = 0, \ldots, g_l(x) = 0, \quad h_1(x) \leq 0, \ldots, h_m(x) \leq 0,$$

$$x_i(x_i - 1) = 0, \quad i = 1, \ldots, n.$$

Assume that the solution set S of the problem is not empty, and define the function $f \colon \mathbb{R}^n \to \mathbb{R}, x \mapsto f(x)$, by

$$f(x) := \max \big\{ |g_1(x)|, \ldots, |g_l(x)|, [h_1(x)]_+, \ldots, [h_m(x)]_+,$$
$$|x_1(x_1 - 1)|, \ldots, |x_n(x_n - 1)| \big\}.$$

It is clear that f is non-negative, continuous semi-algebraic and $S = \{x \in \mathbb{R}^n \mid f(x) = 0\}$. Moreover, f is coercive, i.e. $f(x) \to +\infty$ as $\|x\| \to +\infty$. In particular, f satisfies the Palais–Smale condition at any level $t \geq 0$. By Theorem 3.5, there exist some constants $c > 0, \alpha > 0$, and $\beta > 0$ such that for all $x \in \mathbb{R}^n$, we have

$$c \, \mathrm{dist}(x, S) \leq [f(x)]^\alpha + [f(x)]^\beta.$$

3.6 Goodness at infinity and global Hölderian error bounds

Let $f \colon \mathbb{R}^n \to \mathbb{R}$ be a continuous semi-algebraic function such that $S := \{x \in \mathbb{R}^n \mid f(x) \leq 0\} \neq \emptyset$. In Section 3.4, we have shown that if

f satisfies the Palais–Smale condition at each level $t \geq 0$, then there exist some constants $c > 0, \alpha > 0$, and $\beta > 0$ satisfying the global Hölderian error bound

$$c \operatorname{dist}(x, S) \leq [f(x)]_+^\alpha + [f(x)]_+^\beta \quad \text{for all } x \in \mathbb{R}^n.$$

Theorem 3.6 stated below shows that we can set $\beta = 1$ in this inequality if f has a good asymptotic behavior at infinity.

Definition 3.6. A continuous function $f \colon \mathbb{R}^n \to \mathbb{R}$ is said to be *good at infinity* if there exist some constants $c > 0$ and $R > 0$ such that

$$\mathfrak{m}_f(x) \geq c \quad \text{for all } \|x\| \geq R.$$

Theorem 3.6. *Let $f \colon \mathbb{R}^n \to \mathbb{R}$ be a continuous semi-algebraic function such that $S := \{x \in \mathbb{R}^n \mid f(x) \leq 0\} \neq \emptyset$. Assume that f is good at infinity. Then there exist constants $c > 0$ and $\alpha > 0$ such that*

$$c \operatorname{dist}(x, S) \leq [f(x)]_+^\alpha + [f(x)]_+ \quad \text{for all } x \in \mathbb{R}^n.$$

Proof. By the assumption, there exist some constants $c_1 > 0$ and $R_1 > 0$ such that

$$\mathfrak{m}_f(x) \geq c_1 \quad \text{for all } \|x\| \geq R_1.$$

Fix $s^0 \in S$ and let $c_2 := \frac{c_1}{2} > 0$ and $R_2 := 2R_1 + \|s^0\| > 0$. We first show that

$$c_2 \operatorname{dist}(x, S) \leq [f(x)]_+ \quad \text{for } \|x\| \geq R_2.$$

Suppose, by contradiction, that there is $x^0 \in \mathbb{R}^n$, with $\|x^0\| \geq R_2$, such that

$$c_2 \operatorname{dist}(x^0, S) > [f(x^0)]_+.$$

Then $x^0 \notin S$ and so $f(x^0) > 0$. Note that $\inf_{x \in \mathbb{R}^n}[f(x)]_+ = 0$ since $S \neq \emptyset$ and $[f(x)]_+ \geq 0$ for all $x \in \mathbb{R}^n$. Applying the Ekeland Variational Principle (Theorem 3.1) to the function $\mathbb{R}^n \to \mathbb{R}, x \mapsto [f(x)]_+$, with data $\epsilon := [f(x^0)]_+ = f(x^0) > 0$ and $\lambda := \frac{\operatorname{dist}(x^0, S)}{2} > 0$, we find a point y^0 in $\mathbb{R}^n$ such that the following inequalities hold:

$$[f(y^0)]_+ \le [f(x^0)]_+,$$

$$\|y^0 - x^0\| \le \lambda,$$

$$[f(y^0)]_+ \le [f(x)]_+ + \frac{\epsilon}{\lambda}\|x - y^0\| \quad \text{for all } x \in \mathbb{R}^n.$$

Consequently,

$$\text{dist}(y^0, S) \ge \text{dist}(x^0, S) - \|y^0 - x^0\|$$

$$\ge \text{dist}(x^0, S) - \lambda = \frac{\text{dist}(x^0, S)}{2} > 0,$$

which yields $y^0 \notin S$ and $f(y^0) > 0$. Furthermore, we have

$$\|y^0\| \ge \|x^0\| - \|x^0 - y^0\| \ge \|x^0\| - \lambda = \|x^0\| - \frac{\text{dist}(x^0, S)}{2}$$

$$\ge \|x^0\| - \frac{\|x^0 - s^0\|}{2} \ge \|x^0\| - \frac{\|x^0\| + \|s^0\|}{2} = \frac{\|x^0\| - \|s^0\|}{2}$$

$$\ge \frac{R_2 - \|s^0\|}{2} = R_1.$$

On the other hand, by continuity, f is positive in some open neighborhood of y^0. In particular, for all x near y^0, we have $[f(x)]_+ = f(x)$ and so

$$f(y^0) \le f(x) + \frac{\epsilon}{\lambda}\|x - y^0\|.$$

Hence, y^0 is a local minimizer of the function

$$\mathbb{R}^n \to \mathbb{R}, \quad x \mapsto f(x) + \frac{\epsilon}{\lambda}\|x - y^0\|.$$

Thanks to Lemma 3.1, we deduce successively

$$0 \in \partial\left(f(\cdot) + \frac{\epsilon}{\lambda}\|\cdot - y^0\|\right)(y^0) \subseteq \partial f(y^0) + \frac{\epsilon}{\lambda}\partial\|\cdot - y^0\|(y^0)$$

$$\subseteq \partial f(y^0) + \frac{\epsilon}{\lambda}\overline{\mathbb{B}}.$$

By definition, then

$$\mathfrak{m}_f(y^0) \le \frac{\epsilon}{\lambda} = \frac{2f(x^0)}{\text{dist}(x^0, S)} < 2c_2 = c_1,$$

which is a contradiction. Therefore, we have proved that

$$c_2 \operatorname{dist}(x, S) \le [f(x)]_+ \quad \text{for } \|x\| \ge R_2.$$

On the other hand, it follows from Corollary 1.2 that there exist constants $c_3 > 0$ and $\alpha > 0$ such that

$$c_3 \operatorname{dist}(x, S) \le [f(x)]_+^\alpha \quad \text{for } \|x\| \le R_2.$$

Letting $c := \min\{c_2, c_3\}$ yields the desired result. □

3.7 Newton polyhedra and non-degeneracy conditions

In many problems, the combinatorial information of polynomial maps is important and can be found in their Newton polyhedra. In this section, we recall the definition of Newton polyhedra. Then we introduce the notion of non-degeneracy (at infinity) for polynomial maps in the sense of their Newton polyhedra. As we will see, non-degenerate polynomial maps have a number of remarkable properties which make them an attractive domain for various applications.

Throughout the text, we consider a fixed coordinate system $(x_1, \ldots, x_n) \in \mathbb{R}^n$. Let $J \subset \{1, \ldots, n\}$; then we define

$$\mathbb{R}^J := \{x \in \mathbb{R}^n \mid x_j = 0 \text{ for all } j \notin J\}.$$

We denote by $\mathbb{R}_+$ the set of non-negative real numbers. If $\kappa = (\kappa_1, \ldots, \kappa_n) \in \mathbb{N}^n$, we denote by x^κ the monomial $x_1^{\kappa_1} \cdots x_n^{\kappa_n}$ and by $|\kappa|$ the sum $\kappa_1 + \cdots + \kappa_n$. Note that when $\kappa = (0, \ldots, 0)$, $x^\kappa = 1$.

3.7.1 *Newton polyhedra*

Definition 3.7. A subset $\mathcal{N} \subset \mathbb{R}_+^n$ is said to be a *Newton polyhedron at infinity* if there exists some finite subset $A \subset \mathbb{N}^n$ such that $\mathcal{N}$ is equal to the convex hull in $\mathbb{R}^n$ of $A \cup \{0\}$.[1]

Hence, we say that $\mathcal{N}$ is the Newton polyhedron at infinity determined by A and we write $\mathcal{N} = \mathcal{N}(A)$. A Newton polyhedron at

---

[1] For convenience, we included the origin $0 \in \mathbb{R}^n$ in the definition of $\mathcal{N}$.

infinity $\mathcal{N} \subset \mathbb{R}_+^n$ is said to be *convenient* if it intersects each coordinate axis in a point different from the origin, that is, if for any $i \in \{1, \ldots, n\}$, there exists some integer $m_j > 0$ such that $m_j e_j \in \mathcal{N}$, where the e_j's are the canonical basis vectors in $\mathbb{R}^n$.

Given a Newton polyhedron at infinity $\mathcal{N} \subset \mathbb{R}_+^n$ and a vector $q \in \mathbb{R}^n$, we define

$$d(q, \mathcal{N}) := \min\{\langle q, \kappa \rangle \mid \kappa \in \mathcal{N}\},$$

$$\Delta(q, \mathcal{N}) := \{\kappa \in \mathcal{N} \mid \langle q, \kappa \rangle = d(q, \mathcal{N})\}.$$

We say that a subset Δ of $\mathcal{N}$ is a *face* of $\mathcal{N}$ if there exists a vector $q \in \mathbb{R}^n$ such that $\Delta = \Delta(q, \mathcal{N})$. The dimension of a face Δ is defined as the minimum of the dimensions of the affine subspaces containing Δ. The faces of $\mathcal{N}$ of dimension 0 are called the *vertices* of $\mathcal{N}$. We denote by $\mathcal{N}_\infty$ the set of faces of $\mathcal{N}$ which do not contain the origin 0 in $\mathbb{R}^n$.

Let $\mathcal{N}_1, \ldots, \mathcal{N}_p$ be a collection of p Newton polyhedra at infinity in $\mathbb{R}_+^n$, for some $p \geq 1$. The *Minkowski sum* of $\mathcal{N}_1, \ldots, \mathcal{N}_p$ is defined as the set

$$\mathcal{N}_1 + \cdots + \mathcal{N}_p = \{\kappa^1 + \cdots + \kappa^p \mid \kappa^i \in \mathcal{N}_i \quad \text{for all } i = 1, \ldots, p\}.$$

By definition, $\mathcal{N}_1 + \cdots + \mathcal{N}_p$ is again a Newton polyhedron at infinity. Moreover, we have for all $q \in \mathbb{R}^n$,

$$\begin{aligned} d(q, \mathcal{N}_1 + \cdots + \mathcal{N}_p) &= d(q, \mathcal{N}_1) + \cdots + d(q, \mathcal{N}_p), \\ \Delta(q, \mathcal{N}_1 + \cdots + \mathcal{N}_p) &= \Delta(q, \mathcal{N}_1) + \cdots + \Delta(q, \mathcal{N}_p). \end{aligned} \qquad (3.2)$$

The following result will be useful in the sequel.

Lemma 3.6. *The following two statements are valid:*

(i) *Assume that $\mathcal{N}$ is a convenient Newton polyhedron at infinity. Let Δ be a face of $\mathcal{N}$ and let $q = (q_1, \ldots, q_n) \in \mathbb{R}^n$ be such that $\Delta = \Delta(q, \mathcal{N})$. Then the following conditions are equivalent:*

    (i1) $\Delta \in \mathcal{N}_\infty$.
    (i2) $d(q, \mathcal{N}) < 0$.
    (i3) $\min_{j=1,\ldots,n} q_j < 0$.

(ii) *Assume that $\mathcal{N}_1, \ldots, \mathcal{N}_p$ are some Newton polyhedra at infinity. Let Δ be a face of the Minkowski sum $\mathcal{N} := \mathcal{N}_1 + \cdots + \mathcal{N}_p$. Then the following statements hold:*

(ii1) *There exists a unique collection of faces $\Delta_1, \ldots, \Delta_p$ of $\mathcal{N}_1, \ldots, \mathcal{N}_p$, respectively, such that*

$$\Delta = \Delta_1 + \cdots + \Delta_p.$$

(ii2) *If $\mathcal{N}_1, \ldots, \mathcal{N}_p$ are convenient, then $\mathcal{N}_\infty \subset \mathcal{N}_{1,\infty} + \cdots + \mathcal{N}_{p,\infty}$.*

Proof. (i) First of all, suppose that $\Delta \in \mathcal{N}_\infty$. By definition, $\Delta(q, \mathcal{N}) = \{\kappa \in \mathcal{N} \mid \langle q, \kappa \rangle = d(q, \mathcal{N})\}$ and $\langle q, \kappa \rangle > d(q, \mathcal{N})$ for $\kappa \in \mathcal{N} \setminus \Delta(q, \mathcal{N})$. Since $0 \notin \Delta = \Delta(q, \mathcal{N})$, it follows that $0 = \langle q, 0 \rangle > d(q, \mathcal{N})$. Let $\kappa \in \Delta$, then $\langle q, \kappa \rangle = d(q, \mathcal{N}) < 0$. Note that all coordinates of κ are not negative, so at least one of the coordinates of q must be negative.

Now assume that $d(q, \mathcal{N}) < 0$. By contradiction, suppose that $\Delta \notin \mathcal{N}_\infty$. Hence, $0 \in \Delta$, so by definition, $d(q, \mathcal{N}) = 0$. This is a contradiction.

Finally, assume that $q_{j_*} := \min_{j=1,\ldots,n} q_j < 0$. Since $\mathcal{N}$ is convenient, there exists an integer $m_{j_*} > 0$ such that $m_{j_*} e_{j_*} \in \mathcal{N}$. So $\langle q, e_{j_*} \rangle = q_{j_*} \cdot m_{j_*} < 0$, which implies that $d(q, \mathcal{N}) < 0$.

(ii1) By definition and (3.2), there exists a vector $q \in \mathbb{R}^n$ such that

$$\Delta = \Delta(q, \mathcal{N}_1 + \cdots + \mathcal{N}_p) = \Delta(q, \mathcal{N}_1) + \cdots + \Delta(q, \mathcal{N}_p).$$

It is clear that $\Delta_i := \Delta(q, \mathcal{N}_i)$ is a face of $\mathcal{N}_i$ for $i = 1, \ldots, p$.

(ii2) Let $\Delta \in \mathcal{N}_\infty$ and let $q = (q_1, \ldots, q_n) \in \mathbb{R}^n$ be such that $\Delta = \Delta(q, \mathcal{N})$. Then by (i), $\min_{j=1,\ldots,n} q_j < 0$. By (3.2), $\Delta = \Delta(q, \mathcal{N}_1) + \cdots + \Delta(q, \mathcal{N}_p)$. We need to prove that $\Delta(q, \mathcal{N}_i) \in \mathcal{N}_{i,\infty}$ for $i = 1, \ldots, p$. By contradiction, suppose that there exists an index i_0 such that $\Delta(q, \mathcal{N}_{i_0}) \notin \mathcal{N}_{i_0,\infty}$. Hence, $0 \in \Delta(q, \mathcal{N}_{i_0})$, so by definition, $d(q, \mathcal{N}_{i_0}) = 0$. On the other side, by (i) and by the fact that $\min_{j=1,\ldots,n} q_j < 0$, it follows that $d(q, \mathcal{N}_{i_0}) < 0$. This contradiction ends the proof of the lemma. $\qquad\square$

3.7.2   Non-degeneracy conditions

Let $f \colon \mathbb{R}^n \to \mathbb{R}$ be a polynomial function. Suppose that f is written as $f = \sum_\kappa a_\kappa x^\kappa$. Then the support of f, denoted by $\operatorname{supp}(f)$, is defined as the set of those $\kappa \in \mathbb{N}^n$ such that $a_\kappa \neq 0$. We denote the set $\mathcal{N}(\operatorname{supp}(f))$ by $\mathcal{N}(f)$. This set is called the *Newton polyhedron at infinity* of f. The polynomial f is said to be *convenient* if $\mathcal{N}(f)$ is convenient. If $f \equiv 0$, then we set $\mathcal{N}(f) = \emptyset$. Note that if f is convenient, then for each non-empty subset J of $\{1, \ldots, n\}$, we have $\mathcal{N}(f) \cap \mathbb{R}^J = \mathcal{N}(f|_{\mathbb{R}^J})$. The *Newton boundary at infinity* of f, denoted by $\mathcal{N}_\infty(f)$, is defined as the set of the faces of $\mathcal{N}(f)$ which do not contain the origin 0 in $\mathbb{R}^n$. For each face Δ of $\mathcal{N}_\infty(f)$, we define the *principal part of f at infinity with respect to* Δ, denoted by f_Δ, as the sum of the terms $a_\kappa x^\kappa$ such that $\kappa \in \Delta$.

Remark 3.3. By definition, for each face Δ of $\mathcal{N}_\infty(f)$, there exists a vector $q = (q_1, \ldots, q_n) \in \mathbb{R}^n$ with $\min_{j=1,\ldots,n} q_j < 0$ such that $\Delta = \Delta(q, \mathcal{N})$.

Let $F := (f_1, \ldots, f_p) \colon \mathbb{R}^n \to \mathbb{R}^p, 1 \le p \le n$, be a polynomial map. We say that F is *convenient* if all its components f_i are convenient. Let $\mathcal{N}(F)$ denote the Minkowski sum $\mathcal{N}(f_1) + \cdots + \mathcal{N}(f_p)$, and we denote by $\mathcal{N}_\infty(F)$ the set of faces of $\mathcal{N}(F)$ which do not contain the origin 0 in $\mathbb{R}^n$. Let Δ be a face of the $\mathcal{N}(F)$. According to Lemma 3.6, we have the unique decomposition $\Delta = \Delta_1 + \cdots + \Delta_p$, where Δ_i is a face of $\mathcal{N}(f_i)$ for $i = 1, \ldots, p$. We denote by F_Δ the map $(f_{1,\Delta_1}, \ldots, f_{p,\Delta_p}) \colon \mathbb{R}^n \to \mathbb{R}^p$.

Definition 3.8. (i) We say that $F = (f_1, \ldots, f_p)$ is *Khovanskii non-degenerate at infinity* if and only if for any face Δ of $\mathcal{N}_\infty(F)$ and for all $x \in (\mathbb{R} \setminus \{0\})^n \cap F_\Delta^{-1}(0)$, we have

$$\operatorname{rank} \begin{pmatrix} x_1 \dfrac{\partial f_{1,\Delta_1}}{\partial x_1}(x) & \cdots & x_n \dfrac{\partial f_{1,\Delta_1}}{\partial x_n}(x) \\ \vdots & \cdots & \vdots \\ x_1 \dfrac{\partial f_{p,\Delta_p}}{\partial x_1}(x) & \cdots & x_n \dfrac{\partial f_{p,\Delta_p}}{\partial x_n}(x) \end{pmatrix} = p.$$

(ii) We say that $F = (f_1, \ldots, f_p)$ is *non-degenerate at infinity* if and only if for any face Δ of $\mathcal{N}_\infty(F)$ and for all $x \in (\mathbb{R} \setminus \{0\})^n$, we have

$$\text{rank} \begin{pmatrix} x_1 \dfrac{\partial f_{1,\Delta_1}}{\partial x_1}(x) & \cdots & x_n \dfrac{\partial f_{1,\Delta_1}}{\partial x_n}(x) & f_{1,\Delta_1}(x) & 0 & \cdots & 0 \\ x_1 \dfrac{\partial f_{2,\Delta_2}}{\partial x_1}(x) & \cdots & x_n \dfrac{\partial f_{2,\Delta_2}}{\partial x_n}(x) & 0 & f_{2,\Delta_2}(x) & \cdots & 0 \\ \vdots & \cdots & \vdots & \vdots & \vdots & \ddots & \vdots \\ x_1 \dfrac{\partial f_{p,\Delta_p}}{\partial x_1}(x) & \cdots & x_n \dfrac{\partial f_{p,\Delta_p}}{\partial x_n}(x) & 0 & 0 & \cdots & f_{p,\Delta_p}(x) \end{pmatrix} = p.$$

Remark 3.4. Compared to $F = (f_1, \ldots, f_p)$, the polynomial map

$$F_\Delta : \mathbb{R}^n \to \mathbb{R}^p, \quad x \mapsto F_\Delta(x) := (f_{1,\Delta_1}(x), \ldots, f_{p,\Delta_p}(x)),$$

has the following two remarkable properties:

- The *sparsity*, which means that the number of monomials in f_{i,Δ_i} is much less than that in f_i. Hence, working with F_Δ is easier than with F.
- F_Δ is *quasi-homogeneous*, i.e. there exists a vector $q \in \mathbb{R}^n$, with $\min_j q_j < 0$, such that for each $i = 1, \ldots, p$, we have $\Delta_i = \Delta_i(q, \mathcal{N}(f_i))$ and

$$f_{i,\Delta_i}(t^{q_1} x_1, \ldots, t^{q_n} x_n) = t^{d_i} f_{i,\Delta_i}(x_1, \ldots, x_n)$$

for all $t \in \mathbb{R}$ and all $(x_1, \ldots, x_n) \in \mathbb{R}^n$, where $d_i := d(q, \mathcal{N}(f_i))$.

These facts, in many contexts, allow us to check easily the non-degenerate conditions at infinity.

For each subset $I := \{i_1, \ldots, i_q\} \subset \{1, \ldots, p\}$, we define the polynomial map $F_I : \mathbb{R}^n \to \mathbb{R}^q$ by $F_I(x) = (f_{i_1}(x), \ldots, f_{i_q}(x))$. A connection between non-degeneracy conditions is given by the following statement.

Lemma 3.7. *Let* $F := (f_1, \ldots, f_p) : \mathbb{R}^n \to \mathbb{R}^p, 1 \leq p \leq n$, *be a polynomial map. Then* F *is non-degenerate at infinity if and only if* F_I *is Khovanskii non-degenerate at infinity for all subset* $I \subset \{1, \ldots, p\}$.

Proof. The proof is straightforward and we leave the details to the reader. $\square$

Remark 3.5. (i) The above lemma implies that if F is non-degenerate at infinity then F is Khovanskii non-degenerate at infinity. The converse does not hold in general. Also note that if $p = 1$, then these non-degenerate conditions are equivalent to each other and to the so-called Kouchnirenko non-degenerate condition, which will be studied in Chapter 5.

(ii) We should emphasize that the class of polynomial maps (with fixed Newton polyhedra), which are non-degenerate at infinity, is an open and dense semi-algebraic set in the corresponding Euclidean space of data. For more details, see Theorem 5.2 in Chapter 5.

3.8   Non-degeneracy and global Hölderian error bounds

The next result shows the existence of global Hölderian error bounds with explicitly calculated exponents for systems of polynomials, which are convenient and non-degenerate at infinity.

Theorem 3.7. *Let* $F := (f_1, \ldots, f_p) \colon \mathbb{R}^n \to \mathbb{R}^p, 1 \leq p \leq n,$ *be a polynomial map, convenient and non-degenerate at infinity. Let* $f(x) := \max_{i=1,\ldots,p} f_i(x)$ *and assume* $S := \{x \in \mathbb{R}^n \mid f(x) \leq 0\} \neq \emptyset.$ *Then there exists a constant* $c > 0$ *such that*

$$c \operatorname{dist}(x, S) \leq [f(x)]_+^{\frac{1}{\mathscr{R}(n+p-1,d+1)}} + [f(x)]_+ \quad \textit{for all } x \in \mathbb{R}^n,$$

where $d := \max_{i=1,\ldots,p} \deg f_i.$

The following lemma is crucially used in the proof of Theorem 3.7.

Lemma 3.8. *Under the assumptions of Theorem 3.7,* f *is good at infinity, i.e. there exist some constants* $c > 0$ *and* $R > 0$ *such that*

$$\mathfrak{m}_f(x) \geq c \quad \textit{for all } \|x\| \geq R.$$

Proof. Suppose, by contradiction, that there exists a sequence $\{x^k\}_{k \in \mathbb{N}} \subset \mathbb{R}^n$ such that

$$\lim_{k \to \infty} \|x^k\| = +\infty \quad \text{and} \quad \lim_{k \to \infty} \mathfrak{m}_f(x^k) = 0.$$

By Lemma 3.2, there exists a sequence $\lambda^k := (\lambda_1^k, \ldots, \lambda_p^k) \in \mathbb{R}^p$, with $\lambda_i^k \geq 0$ and $\sum_{i \in I(x^k)} \lambda_i^k = 1$, such that

$$\mathfrak{m}_f(x^k) = \left\| \sum_{i \in I(x^k)} \lambda_i^k \nabla f_i(x^k) \right\|.$$

Recall that $I(x) := \{i \mid f_i(x) = f(x)\}$. Since the number of subsets of $\{1, \ldots, p\}$ is finite, by taking subsequences if necessary, we may assume that the sequence of sets $I(x^k)$ is stable, i.e. there exists $I_1 \subseteq \{1, \ldots, p\}$ such that $I_1 = I(x^k)$ for all k. In view of the Tarski–Seidenberg theorem (Theorem 1.5), the function $\mathfrak{m}_f(x)$ is semi-algebraic. Applying the Curve Selection Lemma at infinity (Theorem 1.12) to the semi-algebraic function

$$A \to \mathbb{R}, \quad (x, \lambda) \mapsto \left\| \sum_{i \in I_1} \lambda_i \nabla f_i(x) \right\|,$$

where

$$A := \left\{ (x, \lambda) \in \mathbb{R}^n \times \mathbb{R}^{\#I_1} \mid f_i(x) = f(x) \text{ for } i \in I_1, \right.$$

$$f_i(x) < f(x) \text{ for } i \notin I_1,$$

$$\left. \lambda_i \geq 0, \ \sum_{i \in I_1} \lambda_i = 1, \ \mathfrak{m}_f(x) = \left\| \sum_{i \in I_1} \lambda_i \nabla f_i(x) \right\| \right\},$$

we find a smooth semi-algebraic curve $\varphi(t) := (\varphi_1(t), \ldots, \varphi_n(t))$ and some smooth semi-algebraic functions $\lambda_i(t), i \in I_1$, for $0 < t \ll 1$, such that

(a) $\lim_{t \to 0} \|\varphi(t)\| = +\infty$;
(b) $f_i(\varphi(t)) = f(\varphi(t))$ for $i \in I_1$, and $f_i(\varphi(t)) < f(\varphi(t))$ for $i \notin I_1$;
(c) $\lambda_i(t) \geq 0$ for all $i \in I_1$, and $\sum_{i \in I_1} \lambda_i(t) = 1$;
(d) $\mathfrak{m}_f(\varphi(t)) = \|\sum_{i \in I_1} \lambda_i(t) \nabla f_i(\varphi(t))\| \to 0$ as $t \to 0^+$.

Let $J := \{j \mid \varphi_j \not\equiv 0\}$. By condition (a), $J \neq \emptyset$. In view of the Growth Dichotomy Lemma (Lemma 1.7), for $j \in J$, we can expand the coordinate φ_j in terms of the parameter, say

$$\varphi_j(t) = x_j^0 t^{q_j} + \text{higher order terms in } t,$$

where $x_j^0 \neq 0$ and $q_j \in \mathbb{Q}$. From condition (a), we get $q_{j_*} := \min_{j \in J} q_j < 0$ for some $j_* \in J$. Note that $\|\varphi(t)\| = c t^{q_{j_*}} + o(t^{q_{j_*}})$ as $t \to 0^+$, for some $c > 0$.

Since f_i is convenient, $\mathcal{N}(f_i) \cap \mathbb{R}^J \neq \emptyset$. Let d_i be the minimal value of the linear function $\sum_{j \in J} q_j \kappa_j$ on $\mathcal{N}(f_i) \cap \mathbb{R}^J$, and let Δ_i be the (unique) maximal face of $\mathcal{N}(f_i) \cap \mathbb{R}^J$ where the linear function takes this value. Since f_i is convenient, $d_i < 0$ and Δ_i is a face of $\mathcal{N}_\infty(f_i)$. Note that f_{i,Δ_i} is not dependent on x_j for all $j \notin J$. By a direct calculation,

$$f_i(\varphi(t)) = f_{i,\Delta_i}(x^0) t^{d_i} + \text{higher order terms in } t,$$

where $x^0 := (x_1^0, \ldots, x_n^0)$ with $x_j^0 := 1$ for $j \notin J$.

Let $I_2 := \{i \in I_1 \,|\, \lambda_i \not\equiv 0\}$. It follows from condition (c) that $I_2 \neq \emptyset$. For $i \in I_2$, we expand the coordinate λ_i in terms of the parameter, say

$$\lambda_i(t) = \lambda_i^0 t^{\theta_i} + \text{higher order terms in } t,$$

where $\lambda_i^0 \neq 0$ and $\theta_i \in \mathbb{Q}$.

For $i \in I_2$ and $j \in J$, we have

$$\frac{\partial f_i}{\partial x_j}(\varphi(t)) = \frac{\partial f_{i,\Delta_i}}{\partial x_j}(x^0) t^{d_i - q_j} + \text{higher order terms in } t.$$

This implies that

$$\sum_{i \in I_2} \lambda_i(t) \frac{\partial f_i}{\partial x_j}(\varphi(t))$$

$$= \sum_{i \in I_2} \left(\lambda_i^0 \frac{\partial f_{i,\Delta_i}}{\partial x_j}(x^0) t^{d_i + \theta_i - q_j} + \text{higher order terms in } t \right)$$

$$= \left(\sum_{i \in I_3} \lambda_i^0 \frac{\partial f_{i,\Delta_i}}{\partial x_j}(x^0) \right) t^{\ell - q_j} + \text{higher order terms in } t,$$

where $\ell := \min_{i \in I_2}(d_i + \theta_i)$ and $I_3 := \{i \in I_2 \,|\, d_i + \theta_i = \ell\} \neq \emptyset$.

There are two cases to be considered.

Case 1: $\ell \leq q_{j_*} := \min_{j \in J} q_j$.

We deduce from condition (d) that

$$\sum_{i \in I_3} \lambda_i^0 \frac{\partial f_{i,\Delta_i}}{\partial x_j}(x^0) = 0 \quad \text{for all } j \in J,$$

which yields

$$\sum_{i \in I_3} \lambda_i^0 \frac{\partial f_{i,\Delta_i}}{\partial x_j}(x^0) = 0 \quad \text{for all } j = 1, 2, \ldots, n$$

because f_{i,Δ_i} is not dependent on x_j for all $j \notin J$. Consequently, we obtain

$$\text{rank}\left(x_j^0 \frac{\partial f_{i,\Delta_i}}{\partial x_j}(x^0) \right)_{i \in I_3, 1 \leq j \leq n} < \#I_3.$$

Since the map $F = (f_1, \ldots, f_p)$ is non-degenerate at infinity, there exists an index $i_0 \in I_3$ such that $f_{i_0,\Delta_{i_0}}(x^0) \neq 0$. Then, by condition (b), we have for all $i \in I_1$,

$$f(\varphi(t)) = f_i(\varphi(t))$$

$$= f_{i_0}(\varphi(t)) = f_{i_0,\Delta_{i_0}}(x^0) t^{d_{i_0}} + \text{higher order terms in } t.$$

By taking the derivative in t of the function $(f \circ \varphi)(t)$, we deduce that

$$\frac{d(f \circ \varphi)(t)}{dt} = \frac{d(f_i \circ \varphi)(t)}{dt} = \left\langle \nabla f_i(\varphi(t)), \frac{d\varphi(t)}{dt} \right\rangle \quad \text{for all } i \in I_1.$$

By condition (c), then

$$\frac{d(f \circ \varphi)(t)}{dt} = \sum_{i \in I_1} \lambda_i(t) \frac{d(f \circ \varphi)(t)}{dt} = \left\langle \sum_{i \in I_1} \lambda_i(t) \nabla f_i(\varphi(t)), \frac{d\varphi(t)}{dt} \right\rangle.$$

Thus,

$$\left| \frac{d(f \circ \varphi)(t)}{dt} \right| \leq \left\| \sum_{i \in I_1} \lambda_i(t) \nabla f_i(\varphi(t)) \right\| \left\| \frac{d\varphi(t)}{dt} \right\| = \mathfrak{m}_f(\varphi(t)) \left\| \frac{d\varphi(t)}{dt} \right\|,$$

which implies that

$$\mathfrak{m}_f(\varphi(t)) \geq c' t^{d_{i_0} - q_{j_*}} + \text{higher order terms in } t,$$

for some $c' > 0$. But this inequality contradicts condition (d) since we know that

$$d_{i_0} \leq d_{i_0} + \theta_{i_0} = \ell \leq q_{j_*}.$$

Case 2: $\ell > q_{j_*} := \min_{j \in J} q_j$.

It follows from condition (c) that $\theta_i \geq 0$ for all $i \in I_2$ and $\theta_i = 0$ for some $i \in I_2$. Without loss of generality, we may assume that $1 \in I$ and $\theta_1 = 0$. Since f_1 is convenient, for any $j = 1, \ldots, n$, there exists a natural number $m_j \geq 1$ such that $m_j e_j \in \mathcal{N}_\infty(f_1)$, where the e_j's are the canonical basis vectors in $\mathbb{R}^n$. Then, by definition,

$$q_j m_j \geq d_1 \quad \text{for all } j \in J.$$

On the other hand, we have

$$d_1 = d_1 + \theta_1 \geq \min_{i \in I_2}(d_i + \theta_i) = \ell.$$

Therefore,

$$q_{j_*} m_{j_*} \geq d_1 \geq \ell > q_{j_*}.$$

Since $q_{j_*} = \min_{j \in J} q_j < 0$, it implies that $m_{j_*} < 1$, which is a contradiction. $\square$

Now, we are in a position to finish the proof of Theorem 3.7.

Proof of Theorem 3.7. By Lemma 3.8, the continuous semi-algebraic function

$$f : \mathbb{R}^n \to \mathbb{R}, \quad x \mapsto f(x) := \max_{i=1,\ldots,p} f_i(x),$$

is good at infinity. Then the proof follows on the same lines as that of Theorem 3.6, by using the error bound with explicit exponent in Theorem 3.3 instead of the classical Łojasiewicz inequality (Corollary 1.2). $\square$

Example 3.6. (i) The partition problem asks whether a sequence of integers $a_1, \ldots, a_n$ can be partitioned, i.e. whether there exists $x \in \{\pm 1\}^n$ such that $\sum_{j=1}^n a_j x_j = 0$. If the infimum of the polynomial $f := \left(\sum_{j=1}^n a_j x_j\right)^2 + \sum_{j=1}^n (x_j^2 - 1)^2$ on $\mathbb{R}^n$ is equal to 0, a global minimizer is ± 1-valued and thus provides a partition of the sequence.

By definition, the Newton polyhedron at infinity $\mathcal{N}(f)$ of f is the convex hull of the origin and the vertices $(4, 0, \ldots, 0), \ldots, (0, \ldots, 4)$, and hence it is convenient. For any face Δ of $\mathcal{N}_\infty(f)$, the principal part of f at infinity with respect to Δ is of the form

$$f_\Delta = \sum_{j \in J} x_j^4,$$

for some non-empty subset $J \subset \{1, \ldots, n\}$. It is clear that

$$\operatorname{rank}\left(x_1 \frac{\partial f_\Delta}{\partial x_1}(x), \ldots, x_n \frac{\partial f_\Delta}{\partial x_n}(x)\right) = 1$$

for all $x \in (\mathbb{R} \setminus \{0\})^n$. So the condition of non-degeneracy at infinity is satisfied.

We have $d := \deg f = 4$ and $p = 1$, so $\mathscr{R}(n + p - 1, d + 1) = \mathscr{R}(n, 5) = 5 \times (12)^{n-1}$. Assume $\inf_{x \in \mathbb{R}^n} f(x) = 0$ and let S be the set of optimal solutions of the problem, i.e.

$$S := \{x \in \mathbb{R}^n \mid f(x) = 0\}.$$

Note that f is coercive, and so f attains its infimum on $\mathbb{R}^n$. In particular, the set S is non-empty. By Theorem 3.7, the following global Hölderian error bound holds:

$$c \operatorname{dist}(x, S) \leq [f(x)]^{\frac{1}{5 \times (12)^{n-1}}} + f(x) \quad \text{for all } x \in \mathbb{R}^n.$$

(ii) Let $d_j, j = 1, \ldots, n$, be positive integers with $d_j \geq 2$ and let

$$f_i(x) := a_{i,1} x_1^{d_1} + \cdots + a_{i,n} x_n^{d_n}, \quad i = 1, \ldots, p \ (p \leq n),$$

be Brieskorn–Pham type polynomials, where $a_{i,j}$ are real numbers satisfying

$$\det((a_{i,j})_{i \in I, j \in J}) \neq 0$$

for any subsets $I \subset \{1, \ldots, p\}$ and $J \subset \{1, \ldots, n\}$ with $\#I = \#J$. It is easy to verify that the map $F := (f_1, \ldots, f_p) \colon \mathbb{R}^n \to \mathbb{R}^p$ is convenient and that the set $S := \{x \in \mathbb{R}^n \mid f_1(x) \leq 0, \ldots, f_p(x) \leq 0\}$ is non-empty because it contains the origin $0 \in \mathbb{R}^n$. Furthermore, for any face $\Delta \in \mathcal{N}_\infty(F)$, there exist faces $\Delta_i \in \mathcal{N}_\infty(f_i)$, with $\Delta = \Delta_1 + \cdots + \Delta_p$, and a subset $J \subset \{1, \ldots, n\}$ such that

$$f_{i,\Delta_i}(x) := \sum_{j \in J} a_{i,j} x_j^{d_j}, \quad i = 1, \ldots, p.$$

This implies that F is non-degenerate at infinity. By Theorem 3.7, there exists a constant $c > 0$ such that

$$c \operatorname{dist}(x, S) \leq [f(x)]_+^{\frac{1}{\mathscr{R}(n+p-1,d+1)}} + [f(x)]_+ \quad \text{for all } x \in \mathbb{R}^n,$$

where $f(x) := \max_{i=1,\ldots,p} f_i(x)$ and $d := \max_{i=1,\ldots,p} d_i$.

3.9 Global Hölderian error bounds for one quadratic inequality

In this section, we establish global Hölderian error bounds with explicit exponent for one quadratic inequality. To do this, let $f(x) := \frac{1}{2} x^T A x + x^T b + c$ be a quadratic polynomial on $\mathbb{R}^n$, where A is a non-zero symmetric $n \times n$-matrix, b is a vector in $\mathbb{R}^n$ and c is a real number. It is well known that A has n real eigenvalues $\lambda_i, i = 1, \ldots, n$. Then we make use of the following notation:

$$\lambda(A) := \min\{|\lambda_i| \, | \, \lambda_i \neq 0, i = 1, \ldots, n\}.$$

Lemma 3.9. *Let $\bar{x} \in \mathbb{R}^n$ be such that $\nabla f(\bar{x}) = 0$. Then for all $x \in \mathbb{R}^n$, the following gradient inequality holds:*

$$\sqrt{2\lambda(A)}|f(x) - f(\bar{x})|^{\frac{1}{2}} \leq \|\nabla f(x)\|.$$

Proof. Since $0 = \nabla f(\bar{x}) = A\bar{x} + b$, $b = -A\bar{x}$. Hence, $\nabla f(x) = Ax + b = A(x - \bar{x})$ and

$$\begin{aligned}
f(x) - f(\bar{x}) &= \tfrac{1}{2}(x^T A x - \bar{x}^T A \bar{x}) + (x - \bar{x})^T b \\
&= \tfrac{1}{2}(x^T A x - \bar{x}^T A \bar{x}) - (x - \bar{x})^T A\bar{x} \\
&= \tfrac{1}{2}(x - \bar{x})^T A(x - \bar{x}).
\end{aligned}$$

On the other hand, since the matrix A is symmetric, it has n real eigenvalues $\lambda_i, i = 1, \ldots, n$. By renumbering, we can assume that $\lambda_i, i = 1, \ldots, \kappa$, are the non-zero eigenvalues of A for some integer $\kappa \leq n$. Then we can write

$$A = M \text{ diag}(\lambda_1, \ldots, \lambda_\kappa, 0, \ldots, 0) M^T,$$

where M is a suitable orthonormal $n \times n$ matrix.

Now, let $x \in \mathbb{R}^n$ and set $z := M^T(x - \bar{x}) \in \mathbb{R}^n$. We have

$$\|\nabla f(x)\|^2 = \|Ax + b\|^2 = \|A(x - \bar{x})\|^2 = \sum_{i=1}^\kappa \lambda_i^2 z_i^2$$

and

$$|f(x) - f(\bar{x})| = \frac{1}{2} \left| (x - \bar{x})^T A(x - \bar{x}) \right|$$

$$= \frac{1}{2} \left| \sum_{i=1}^\kappa \lambda_i z_i^2 \right| \leq \frac{1}{2} \sum_{i=1}^\kappa |\lambda_i| z_i^2,$$

which implies the gradient inequality. $\qquad\square$

As a consequence of Lemma 3.9, we have the following global error bound for one quadratic inequality.

Theorem 3.8. *Let $S := \{x \in \mathbb{R}^n | f(x) \leq 0\}$ and assume that $S \neq \emptyset$. There exists some constant $c > 0$ such that for all $x \in \mathbb{R}^n$,*

$$c \, \text{dist}(x, S) \leq [f(x)]_+^\alpha,$$

where the exponent α is either 1 or $\frac{1}{2}$.

Proof. By performing an invertible affine transformation if necessary, we can assume f is in the following standard form:

$$f(x) = \sum_{i \in I_+} x_i^2 - \sum_{i \in I_-} x_i^2 + \sum_{i \in J} x_i,$$

where I_+, I_- and J are some non-overlapping subsets of $\{1, 2, \ldots, n\}$. Let $\bar{x} := 0$. At this point, the proof breaks up into two cases:

Case 1: $J = \emptyset$.

In this situation, we have $f(\bar{x}) = 0$ and $\nabla f(\bar{x}) = 0$. By Lemma 3.9,

$$\sqrt{2} |f(x)|^{\frac{1}{2}} \leq \|\nabla f(x)\| \quad \text{for all } x \in \mathbb{R}^n,$$

This, together with Lemma 3.3, implies that

$$\tfrac{\sqrt{2}}{2}\,\mathrm{dist}(x,S) \leq [f(x)]_+^{\frac{1}{2}} \quad \text{for all } x \in \mathbb{R}^n,$$

which proves the theorem in Case 1.

Case 2: $J \neq \emptyset$.

In this situation, we have $\|\nabla f(x)\| \geq \sum_{j \in J} 1 \geq 1$ for all $x \in \mathbb{R}^n$. Thanks to Lemma 3.3, then

$$\tfrac{1}{2}\,\mathrm{dist}(x,S) \leq [f(x)]_+ \quad \text{for all } x \in \mathbb{R}^n,$$

which proves the theorem in Case 2. □

Bibliographic notes

The results in Sections 3.2 and 3.3 are taken from the paper by Li, Mordukhovich and Phạm [90]. The materials in Sections 3.4–3.8 are based on the work of Đinh, Hà and Phạm [29] which generalized the results of Hà [54]. For more information concerning error bounds with explicit exponents, we refer the reader to the papers in [25–27, 30, 31, 91, 127].

The study of error bounds has attracted a lot of attention of many researchers over the years and has found numerous applications to, in particular, sensitivity analysis for various problems of mathematical programming, termination criteria for numerical algorithms, etc. The fundamental results were obtained in the works of Hoffman [58], Robinson [133], Mangasarian [97], Auslender and Crouzeix [7], Klatte [67], Klatte and Li [68] and others [86, 88, 89, 108, 109, 147]. The reader is referred to the survey paper by Pang [123] for the relevant work and the references therein for the theory and applications of error bounds.

For non-smooth analysis, we refer to the texts in [103, 134]; the results in Section 3.1 can be found in these references. The Variational Principle (Theorem 3.1) was established in 1972 by Ekeland [37]. We have followed the proof of [57]. Lemma 3.3 can be found in [109].

Explicit estimates for the Łojasiewicz exponent in error bounds for real algebraic sets are established by Kurdyka and Spodzieja [74] and independently by Phạm [126] (see also [46, 70]). Example 3.2 is taken from [70].

Lemma 3.9 is given in [38, Property 6]. Theorem 3.8 is taken from the papers in [96, 108]; the proof given here is different from the originals.

0/1 integer feasibility problem is considered in [94]. The partition problem, which is known to be NP-complete, can be found in [42].

The Palais–Smale condition originally introduced in [118] and the Palais–Smale condition at the level t is introduced by Brézis, Coron and Nirenberg in [18].

The conditions of non-degeneracy for complex analytic maps $F \colon (\mathbb{C}^n, 0) \to (\mathbb{C}^p, 0)$ are introduced by Kouchnirenko [71] for $p = 1$ and by Khovanskii [66] for $p \geq 1$. These are very important notions in Theory of Singularities and Algebraic Geometry. They allow us to control the behavior of the gradient of polynomial maps in neighborhoods of singular points or in a neighborhood of a point at infinity. We refer the reader to [5] for many applications of these conditions. The proofs of genericity for non-degenerate maps can be found in [26, 71, 117].

The notion of goodness at infinity is inspired by Broughton [17], who called a complex polynomial $f \colon \mathbb{C}^n \to \mathbb{C}$ *tame* if there exist $c > 0$ and $R > 0$ such that

$$\|\nabla f(x)\| \geq c \quad \text{for all } x \in \mathbb{C}^n, \ \|x\| \geq R.$$

He proved that a complex polynomial is tame provided that it is convenient and non-degenerate at infinity. The proof of Lemma 3.8 presented here is different from that given in [17, Proposition 3.4].

Chapter 4

Frank–Wolfe Type Theorem for Polynomial Programs

Abstract

We deal with the question whether a polynomial optimization problem has an optimal solution or not. More precisely, let $f_0, f_1, \ldots, f_p$ be polynomial functions on $\mathbb{R}^n$ and consider the following constrained optimization problem:

$$\inf f_0(x) \quad \text{subject to} \quad f_1(x) \leq 0, \ldots, f_p(x) \leq 0.$$

In the case when all $f_i, i = 0, \ldots, p$, are affine linear, it is well known that the set of optimal solutions is non-empty, provided that the problem is bounded from below. In 1956, Frank and Wolfe proved that if f_i remains affine linear for $i = 1, \ldots, p$, and f_0 is an arbitrary quadratic polynomial, then the condition of f_0 being bounded from below over the feasible region implies that an optimal solution exists. In this chapter, we will show that a similar statement holds when all the polynomials f_i involved are either convex or convenient and non-degenerate at infinity.

4.1 Frank–Wolfe type theorem for convex polynomial programs

The existence of optimal solutions is an important issue in optimization theory. In this section, we will establish a Frank–Wolfe type theorem for convex polynomial programs. We begin with some properties of convex polynomials.

4.1.1 *Some properties of convex polynomials*

Recall that a real-valued function $f\colon \mathbb{R}^n \to \mathbb{R}$ is called *convex* if for all $x, y \in \mathbb{R}^n$ and all $t \in [0, 1]$,

$$f(tx + (1 - t)y) \le tf(x) + (1 - t)f(y).$$

Lemma 4.1. *Let $f\colon \mathbb{R}^n \to \mathbb{R}$ be a convex polynomial, and let $x, y, v \in \mathbb{R}^n$. If $\phi(t) := f(x + tv)$ defined on $\mathbb{R}$ is a convex polynomial of degree d, then $\psi(t) := f(y + tv), t \in \mathbb{R}$, is also a convex polynomial of degree d. Furthermore, if $d \ge 1$, then the coefficients associated with the terms t^d in $\phi(t)$ and $\psi(t)$ are identical.*

Proof. In what follows, we will need only the case $d \le 1$, and so we restrict the proof to $d \le 1$. However, the proof for arbitrary d is similar, and we leave it to the reader as an exercise.

Let $t \ne 0$ be arbitrary and let z be the point $z = x + 2(y - x)$, i.e. $y = \frac{1}{2}x + \frac{1}{2}z$. Since f is convex and $d \le 1$, there exist real numbers ρ and α such that

$$f(y + tv) = f\left(\tfrac{1}{2}z + \tfrac{1}{2}(x + 2tv)\right)$$
$$\le \tfrac{1}{2}f(z) + \tfrac{1}{2}\phi(2t) = \rho + \alpha t. \tag{4.1}$$

If $\alpha = 0$ (i.e. in the case $d = 0$), then $\psi(t) = f(y + tv)$ is constant because a non-constant convex polynomial in one variable is not bounded from above. We now assume that $d = 1$. Since (4.1) holds for all t, the polynomial $\psi(t)$ is not constant. If $\psi(t)$ is a convex polynomial of degree $d' \ge 2$, then, in particular, the coefficient γ associated with $t^{d'}$ in $\psi(t)$ is positive. Hence, after dividing the inequality (4.1) by $t^{d'}$ on both sides and taking the limit for $t \to +\infty$, the left-hand side of (4.1) tends to $\gamma > 0$ and the right-hand side of (4.1) tends to zero, a contradiction.

To prove the second statement, let again $d = 1$. Hence, both $\phi(t)$ and $\psi(t)$ have linear representations $\phi(t) = \beta + \alpha t$ and $\psi(t) = b + at$ with the same α as in (4.1) and certain β, a, b. Then (4.1) yields

$$\psi(t) = b + at \le \tfrac{1}{2}f(z) + \tfrac{1}{2}\beta + \alpha t \quad \text{for all } t \in \mathbb{R},$$

which implies easily that $a = \alpha$. $\qquad\square$

Denote by 0^+Z the *recession cone* of a convex set $Z \subset \mathbb{R}^n$, i.e.

$$0^+Z := \{v \in \mathbb{R}^n \mid x + tv \in Z \text{ for all } x \in Z \text{ and } t \geq 0\}.$$

Later on, we will make use of the following simple observation.

Lemma 4.2. *Let $f \colon \mathbb{R}^n \to \mathbb{R}$ be a convex polynomial such that the set*

$$L(f) := \{x \in \mathbb{R}^n \mid f(x) \leq 0\}$$

is non-empty. If $v \in 0^+L(f)$ then $f(tv) = f(0) + \alpha t$ for some $\alpha \leq 0$, where $\alpha = 0$ if $\sup_{t \in \mathbb{R}} f(tv) < \infty$ and $\alpha < 0$ if $\sup_{t \in \mathbb{R}} f(tv) = +\infty$.

Proof. Since $v \in 0^+L(f)$, for each x in $L(f)$, we have

$$f(x + tv) \leq 0 \quad \text{for all } t \geq 0.$$

Fixing x in $L(f)$, then $\phi \colon \mathbb{R} \to \mathbb{R}, t \mapsto f(x+tv)$, defines a convex polynomial in one variable. Hence, ϕ cannot have degree greater than 1, otherwise it is not bounded from above on the ray $\{t \in \mathbb{R} \mid t \geq 0\}$ (note that if $\deg \phi \geq 2$, then the coefficient associated with the term of highest degree is positive). By Lemma 4.1, $\psi \colon \mathbb{R} \to \mathbb{R}, t \mapsto f(tv)$, is also a convex polynomial of the same degree $\deg \phi \leq 1$. Then the second statement easily follows. $\qquad \square$

The next lemma states that the orthogonal projection along a recession direction of a convex polynomial set yields again a convex polynomial set. This will be an essential tool in the proof of the corresponding Frank–Wolfe type theorem.

Lemma 4.3. *Let $f_i \colon \mathbb{R}^n \to \mathbb{R}, i = 1, \ldots, p$, be convex polynomials. Suppose that $v \neq 0$ belongs to the recession cone of the non-empty set*

$$S := \{x \in \mathbb{R}^n \mid f_1(x) \leq 0, \ldots, f_p(x) \leq 0\}.$$

Define

$$I := \left\{ i \;\middle|\; \sup_{t \in \mathbb{R}} f_i(tv) < +\infty \right\}.$$

Then the orthogonal projection $\mathrm{pr}\, S$ *of* S *to the subspace* $v^{\perp} := \{x \in \mathbb{R}^n \mid \langle v, x \rangle = 0\}$ *becomes*

$$\mathrm{pr}\, S = \begin{cases} v^{\perp} & \text{if } I = \emptyset, \\ \{x \in \mathbb{R}^n \mid \langle v, x \rangle = 0, f_i(x) \leq 0, i \in I\} & \text{otherwise.} \end{cases}$$

In particular, $\mathrm{pr}\, S$ *is also a convex polynomial set.*

Proof. There are two cases to be considered.

Case 1: $I = \emptyset$.

Since a ray in direction v belongs to S, in the present case, it follows from Lemmas 4.1 and 4.2 that all polynomials $f_i(x), i = 1, \ldots, p$, are linearly decreasing along $\{x + tv \mid t \geq 0\}$. Hence, for any $z \in v^{\perp}$, we have $f_i(z + tv) \leq 0$ for all $i = 1, \ldots, p$, and for sufficiently large t. Therefore, $z \in \mathrm{pr}\, S$.

Case 2: $I \neq \emptyset$.

By Lemmas 4.1 and 4.2, the convex polynomials $f_i, i \in I$, are constant on each $l_x := \{x + tv \mid t \in \mathbb{R}\}$, but the convex polynomials $f_i, i \in \{1, \ldots, p\} \setminus I$, are linearly decreasing on the lines l_x. Hence, for any element z in $Q := \{x \in \mathbb{R}^n \mid \langle v, x \rangle = 0, f_i(x) \leq 0, i \in I\}$, we have $f_i(z + tv) \leq 0$ for all $i = 1, \ldots, p$ and for sufficiently large t. Thus, $z \in \mathrm{pr}\, S$. On the other hand, if z belongs to $v^{\perp}$ but is not an element of Q, then there is some $i_0 \in I$ such that $f_{i_0}(z) > 0$. Therefore, for any real t, we have $f_{i_0}(z + tv) > 0$ (since f_{i_0} is constant), and so, $z \notin \mathrm{pr}\, S$. □

4.1.2   Frank–Wolfe type theorem

The following result is a Frank–Wolfe type theorem for convex polynomial programs.

Theorem 4.1. *Let* $f_i \colon \mathbb{R}^n \to \mathbb{R}, i = 0, 1, \ldots, p$, *be convex polynomials. If* f_0 *is bounded from below on the non-empty set*

$$S := \{x \in \mathbb{R}^n \mid f_1(x) \leq 0, \ldots, f_p(x) \leq 0\},$$

then f_0 *attains its infimum on* S.

Proof. The proof is by induction on the dimension of S.

If $\dim S = 0$, then S is a singleton, and the statement of the theorem trivially holds. Suppose the statement is shown for all convex polynomial feasible sets of dimension $\leq k$, and let now $\dim S = k+1$.

Let M be a real number such that $M \geq f_0(x^0)$ for some $x^0 \in S$. Let

$$S_M := \{x \in \mathbb{R}^n \mid M - f_0(x) \geq 0, f_1(x) \leq 0, \ldots, f_p(x) \leq 0\}.$$

It is clear that the set S_M is convex and closed. By assumption, f_0 is bounded from below on S and hence on S_M. Furthermore, we have

$$\inf_{x \in S} f_0(x) = \inf_{x \in S_M} f_0(x).$$

Therefore, if S_M is bounded, then the Weierstrass theorem yields the desired result.

Now assume that S_M is unbounded. Then, the convexity of S_M implies that S_M contains a ray $r = \{z + tv \mid t \geq 0\}$, i.e. v belongs to the recession cone of S_M. By definition of S_M and by the assumption, the polynomial f_0 is bounded on S_M, hence, f_0 is constant on r according to Lemmas 4.1 and 4.2. Again from Lemma 4.1, it thus follows that f_0 is constant along each ray $\{x + tv \mid t \geq 0\}$ and so, f_0 is also constant along each line $\{x + tv \mid t \in \mathbb{R}\}$ (for arbitrary $x \in \mathbb{R}^n$). Therefore,

$$\inf_{x \in \text{pr } S} f_0(x) = \inf_{x \in S} f_0(x).$$

By Lemma 4.3, $\text{pr } S$ is again a convex polynomial set. By the assumption of induction, there is some $x^* \in \text{pr } S$ such that $\inf_{x \in \text{pr } S} f_0(x) = f_0(x^*)$. Then for sufficiently large positive t we will have $x^* + tv \in S$, and hence,

$$f_0(x^* + tv) = f_0(x^*) = \inf_{x \in \text{pr } S} f_0(x) = \inf_{x \in S} f_0(x).$$

This concludes the proof of the theorem. $\qquad\square$

Example 4.1. Let f_0 and f_1 be the following convex polynomials from $\mathbb{R}^3$ to $\mathbb{R}$:

$$f_0(x, y, z) := (x - y)^4 + z \quad \text{and} \quad f_1(x, y, z) := z^2.$$

Define $S := \{(x, y, z) \in \mathbb{R}^3 \mid f_1(x, y, z) \leq 0\}$. Obviously, f_0 is bounded from below and attains its infimum on S. Furthermore, the set of minimizers of f_0 on S is $\{(x, x, 0) \in \mathbb{R}^3 \mid x \in \mathbb{R}\}$.

Remark 4.1. The convexity of the polynomials $f_0, f_1, \ldots, f_p$ cannot be dropped in the statement of Theorem 4.1. A counterexample is $f_0(x, y) := x^2 + (xy - 1)^2$ and $S := \mathbb{R}^2$. It is easy to check that $\inf_{(x,y) \in S} f_0(x, y) = 0$. However, $f_0(x, y) > 0$ for all $(x, y) \in S$.

4.2 Frank–Wolfe type theorem for non-degenerate polynomial programs

The purpose of this section is to establish a Frank–Wolfe type theorem for non-degenerate polynomial programs. We start with some properties of polynomial maps, which are convenient and non-degenerate at infinity (see Section 3.7).

4.2.1 *Non-degeneracy and the Palais–Smale condition*

Let $F = (f_1, \ldots, f_p) \colon \mathbb{R}^n \to \mathbb{R}^p, 1 \leq p \leq n$, be a polynomial map, and define the *Rabier function* $\nu_F \colon \mathbb{R}^n \to \mathbb{R}$ by

$$\nu_F(x) := \min_{\sum_{i=1}^p |\lambda_i| = 1} \left\| \sum_{i=1}^p \lambda_i \nabla f_i(x) \right\|.$$

Next, we introduce the set of *asymptotic critical values* of F:

$$\widetilde{K}_\infty(F) := \left\{ y \in \mathbb{R}^p \mid \text{there exists } \{x^k\}_{k \in \mathbb{N}} \subset \mathbb{R}^n \text{ such that} \right.$$

$$\left. \lim_{k \to \infty} \|x^k\| = \infty, \ \lim_{k \to \infty} F(x^k) = y, \ \lim_{k \to \infty} \nu_F(x^k) = 0 \right\}.$$

Remark 4.2. (i) By definition, $\nu_F(x) = 0$ if and only if the gradient vectors $\nabla f_1(x), \ldots, \nabla f_p(x)$ are linearly dependent.

(ii) By the Tarski–Seidenberg theorem (Theorem 1.5), we can see that the function ν_F is semi-algebraic and $\widetilde{K}_\infty(F)$ is a semi-algebraic set.

Definition 4.1. We say that the map F satisfies the *Palais–Smale condition* if $\widetilde{K}_\infty(f) = \emptyset$.

The following result will be useful for our later analysis.

Theorem 4.2. *If the map F is convenient and Khovanskii non-degenerate at infinity, then F satisfies the Palais–Smale condition.*

Proof. By contradiction, suppose that $\widetilde{K}_\infty(F) \neq \emptyset$; i.e. there exist a point $y \in \mathbb{R}^p$ and a sequence $\{x^k\}_{k \in \mathbb{N}} \subset \mathbb{R}^n$ such that

$$\lim_{k \to +\infty} \|x^k\| = \infty, \quad \lim_{k \to +\infty} F(x^k) = y, \quad \text{and} \quad \lim_{k \to +\infty} \nu_F(x^k) = 0.$$

It follows that there exists a sequence $\lambda^k := (\lambda_1^k, \ldots, \lambda_p^k) \in \mathbb{R}^p$, with $\sum_{i=1}^p |\lambda_i^k| = 1$, such that we have for all $k \geq 1$,

$$\nu_F(x^k) = \left\| \sum_{i=1}^p \lambda_i^k \nabla f_i(x^k) \right\|.$$

By applying the Curve Selection Lemma at infinity (Theorem 1.12) with the following setup: the set $A := \{(x, \lambda) \in \mathbb{R}^n \times \mathbb{R}^p \mid \sum_{i=1}^p |\lambda_i| = 1, \nu_F(x) = \|\sum_{i=1}^p \lambda_i \nabla f_i(x)\|\}$ which is clearly semi-algebraic, the sequence $(x^k, \lambda^k) \in A$ which tends to infinity as $k \to +\infty$, and the semi-algebraic map $A \to \mathbb{R}^{p+1}, (x, \lambda) \mapsto (F(x), \nu_F(x))$, we find smooth semi-algebraic curves $\varphi(t) := (\varphi_1(t), \ldots, \varphi_n(t))$ and $\lambda(t) := (\lambda_1(t), \ldots, \lambda_p(t)), 0 < t < \epsilon, (\epsilon > 0)$ such that

(a) $\lim_{t \to 0} \|\varphi(t)\| = \infty$;
(b) $\lim_{t \to 0} F(\varphi(t)) = y \in \mathbb{R}^p$;
(c) $\sum_{i=1}^p |\lambda_i(t)| = 1$; and
(d) $\lim_{t \to 0} \|\sum_{i=1}^p \lambda_i(t) \nabla f_i(\varphi(t))\| = 0$.

Let $J := \{j \mid \varphi_j \not\equiv 0\}$. Condition (a) yields $J \neq \emptyset$. By the Growth Dichotomy Lemma (Lemma 1.7), for each $j \in J$, we can expand the coordinate φ_j as follows:

$$\varphi_j(t) = x_j^0 t^{q_j} + \text{higher order terms in } t,$$

where $x_j^0 \neq 0$ and $q_j \in \mathbb{Q}$. From condition (a), we get $\min_{j \in J} q_j < 0$.

Recall that $\mathbb{R}^J := \{\kappa := (\kappa_1, \kappa_2, \ldots, \kappa_n) \in \mathbb{R}^n \mid \kappa_j = 0 \text{ for } j \notin J\}$ and $\mathcal{N}(f_i)$ (respectively, $\mathcal{N}_\infty(f_i)$) is the Newton polyhedron (respectively, the Newton boundary) at infinity of f_i (see Section 3.7).

Since the map $F = (f_1, \ldots, f_p)$ is convenient, $\mathcal{N}(f_i) \cap \mathbb{R}^J \neq \emptyset$ for all $i = 1, \ldots, p$. Let d_i be the minimal value of the linear function $\sum_{j \in J} q_j \kappa_j$ on $\mathcal{N}(f_i) \cap \mathbb{R}^J$, and let Δ_i be the (unique) maximal face of $\mathcal{N}(f_i) \cap \mathbb{R}^J$ where the linear function takes this value. Since f_i is convenient and $\min_{j \in J} q_j < 0$, by Lemma 3.6(i), we have $d_i < 0$ and Δ_i is a face of $\mathcal{N}_\infty(f_i)$. Note that f_{i,Δ_i} does not depend on x_j for all $j \notin J$. Now suppose that f_i is written as $f_i = \sum_\kappa a_{i,\kappa} x^\kappa$. Then

$$
\begin{aligned}
&f_i \circ \varphi(t) \\
&= \sum_{\kappa \in \mathcal{N}(f_i) \cap \mathbb{R}^J} a_{i,\kappa} (\varphi(t))^\kappa \\
&= \sum_{\kappa \in \mathcal{N}(f_i) \cap \mathbb{R}^J} a_{i,\kappa} (\varphi_1(t))^{\kappa_1} \ldots (\varphi_n(t))^{\kappa_n} \\
&= \sum_{\kappa \in \mathcal{N}(f_i) \cap \mathbb{R}^J} \left(a_{i,\kappa} (x_1^0 t^{q_1})^{\kappa_1} \ldots (x_n^0 t^{q_n})^{\kappa_n} + \text{higher order terms in } t \right) \\
&= \sum_{\kappa \in \mathcal{N}(f_i) \cap \mathbb{R}^J} \left(a_{i,\kappa} (x^0)^\kappa t^{\sum_{j \in J} q_j \kappa_j} + \text{higher order terms in } t \right) \\
&= \sum_{\kappa \in \Delta_i} a_{i,\kappa} (x^0)^\kappa t^{d_i} + \text{higher order terms in } t \\
&= f_{i,\Delta_i}(x^0) t^{d_i} + \text{higher order terms in } t,
\end{aligned}
$$

where $x^0 := (x_1^0, \ldots, x_n^0)$ with $x_j^0 := 1$ for $j \notin J$. By condition (b) and $d_i < 0$, we have

$$
f_{i,\Delta_i}(x^0) = 0 \quad \text{for all } i = 1, \ldots, p. \tag{4.2}
$$

Let $I := \{i \mid \lambda_i \not\equiv 0\}$. It follows from condition (c) that $I \neq \emptyset$. For each $i \in I$, we expand the coordinate λ_i in terms of the parameter (cf. Lemma 1.7) as follows

$$
\lambda_i(t) = \lambda_i^0 t^{\theta_i} + \text{higher order terms in } t,
$$

where $\lambda_i^0 \neq 0$ and $\theta_i \in \mathbb{Q}$.

For $i \in I$ and $j \in J$, by some similar calculations as with $f_i(\varphi(t))$, we have

$$\frac{\partial f_i}{\partial x_j}(\varphi(t)) = \frac{\partial f_{i,\Delta_i}}{\partial x_j}(x^0)t^{d_i - q_j} + \text{higher order terms in } t.$$

It implies that

$$\sum_{i \in I} \lambda_i(t) \frac{\partial f_i}{\partial x_j}(\varphi(t))$$

$$= \sum_{i \in I} \left(\lambda_i^0 \frac{\partial f_{i,\Delta_i}}{\partial x_j}(x^0)t^{d_i + \theta_i - q_j} + \text{higher order terms in } t \right)$$

$$= \left(\sum_{i \in I'} \lambda_i^0 \frac{\partial f_{i,\Delta_i}}{\partial x_j}(x^0) \right) t^{\ell - q_j} + \text{higher order terms in } t,$$

where $\ell := \min_{i \in I}(d_i + \theta_i)$ and $I' := \{i \in I \mid d_i + \theta_i = \ell\} \neq \emptyset$. We note, by condition (d), that for all $j \in J$,

$$\left\| \sum_{i \in I} \lambda_i(t) \frac{\partial f_i}{\partial x_j}(\varphi(t)) \right\| = \left\| \sum_{i=1}^{p} \lambda_i(t) \frac{\partial f_i}{\partial x_j}(\varphi(t)) \right\| \to 0, \quad \text{as } t \to 0.$$

At this point, the proof breaks up into two cases.

Case 1: $\ell \leq q_{j_*} := \min_{j \in J} q_j$.

In this situation, we have for all $j \in J$,

$$\sum_{i \in I'} \lambda_i^0 \frac{\partial f_{i,\Delta_i}}{\partial x_j}(x^0) = 0.$$

On the other hand, for each $j \notin J$, the polynomial f_{i,Δ_i} does not depend on the variable x_j, so $\frac{\partial f_{i,\Delta_i}}{\partial x_j} \equiv 0$. Therefore,

$$\sum_{i \in I'} \lambda_i^0 \frac{\partial f_{i,\Delta_i}}{\partial x_j}(x^0) = 0 \quad \text{for all } j = 1, \ldots, n.$$

Consequently, we obtain

$$\text{rank} \begin{pmatrix} x_1^0 \dfrac{\partial f_{1,\Delta_1}}{\partial x_1}(x^0) & \cdots & x_n^0 \dfrac{\partial f_{1,\Delta_1}}{\partial x_n}(x^0) \\ \vdots & \cdots & \vdots \\ x_1^0 \dfrac{\partial f_{p,\Delta_p}}{\partial x_1}(x^0) & \cdots & x_n^0 \dfrac{\partial f_{p,\Delta_p}}{\partial x_n}(x^0) \end{pmatrix} < p$$

since the matrix has $\#I'$ linearly dependent rows. This, together with (4.2), contradicts the assumption that the polynomial map $F = (f_1, \ldots, f_p)$ is Khovanskii non-degenerate at infinity.

Case 2: $\ell > q_{j_*} := \min_{j \in J} q_j$.

It follows from condition (c) that $\theta_i \geq 0$ for all $i \in I$ and $\theta_i = 0$ for some $i \in I$. Without loss of generality, we may assume that $1 \in I$ and $\theta_1 = 0$. Since f_1 is convenient, for any $j = 1, \ldots, n$, there exists a natural number $m_j \geq 1$ such that $m_j e_j \in \mathcal{N}_\infty(f_1)$. Then, by definition, it is clear that

$$q_j m_j \geq d_1 \quad \text{for all } j \in J.$$

On the other hand, we have

$$d_1 = d_1 + \theta_1 \geq \min_{i \in I}(d_i + \theta_i) = \ell.$$

Therefore,

$$q_{j_*} m_{j_*} \geq d_1 \geq \ell > q_{j_*}.$$

Since $q_{j_*} = \min_{j \in J} q_j < 0$, it implies that $m_{j_*} < 1$, which is a contradiction. $\qquad\square$

Corollary 4.1. *Let $F := (f_1, \ldots, f_p) \colon \mathbb{R}^n \to \mathbb{R}^p, 1 \leq p \leq n$, be a polynomial map. If F is convenient and non-degenerate at infinity, then $\widetilde{K}_\infty(F_I) = \emptyset$ for all non-empty subset $I \subset \{1, \ldots, p\}$.*

Proof. The statement follows immediately from Lemma 3.7 and Theorem 4.2. $\qquad\square$

4.2.2 *Non-degeneracy and regularity at infinity*

Let $(f_1, \ldots, f_p) \colon \mathbb{R}^n \to \mathbb{R}^p$ be a polynomial map and assume that

$$S := \{x \in \mathbb{R}^n \mid f_1(x) \le 0, \ldots, f_p(x) \le 0\} \ne \emptyset.$$

For a later result, we will need the following definition.

Definition 4.2. The set S is called *regular at infinity* if there exists a real number $R > 0$ such that for each $x \in S, \|x\| \ge R$, the gradient vectors $\nabla f_i(x), i \in I(x)$, are linearly independent, where $I(x)$ stands for the set of indices i for which f_i vanishes at x.

Remark 4.3. By definition, if S is regular then it is regular at infinity, but the converse does not hold in general.

Lemma 4.4. *Assume that the polynomial map* $(f_1, \ldots, f_p) \colon \mathbb{R}^n \to \mathbb{R}^p$ *is convenient and non-degenerate at infinity. Then the set* $S := \{x \in \mathbb{R}^n \mid f_1(x) \le 0, \ldots, f_p(x) \le 0\}$ *is regular at infinity.*

Proof. Suppose that the lemma does not hold. Then the set

$$S' := \{x \in S \mid I(x) \ne \emptyset,$$

$$\text{the vectors } \nabla f_i(x), i \in I(x), \text{ are linearly dependent}\}$$

is unbounded. Since the number of subsets of $\{1, \ldots, p\}$ is finite, there exists a non-empty subset $I := \{i_1, \ldots, i_q\} \subset \{1, \ldots, p\}$ such that

$$S'' := \{x \in S \mid \text{the gradient vectors } \nabla f_i(x), i \in I,$$

$$\text{are linearly dependent}\}$$

is unbounded. Clearly, the set S'' is semi-algebraic. By the Curve Selection Lemma at infinity (Theorem 1.12), there exists a smooth semi-algebraic curve $\varphi \colon (0, \epsilon) \to \mathbb{R}^n, t \mapsto \varphi(t)$, such that

(a) $\lim_{t \to 0^+} \|\varphi(t)\| = \infty$;
(b) $f_i(\varphi(t)) \equiv 0$ for $i \in I$ and $f_i(\varphi(t)) < 0$ for $i \notin I$;
(c) the gradient vectors $\nabla f_i(\varphi(t)), i \in I$, are linearly dependent.

Therefore, $\nu_{F_I}(\varphi(t)) = 0$ for all $t \in (0, \epsilon)$, where F_I stands for the polynomial map $x \mapsto (f_{i_1}(x), \dots, f_{i_q}(x))$. Consequently, we have $0 \in \widetilde{K}_\infty(F_I)$, which contradicts Corollary 4.1. $\qquad\square$

Lemma 4.5. *Suppose that the closed semi-algebraic set*

$$S := \{x \in \mathbb{R}^n \mid f_1(x) \leq 0, \dots, f_p(x) \leq 0\}$$

is unbounded and regular at infinity. Then there exists a real number $R_0 > 0$ such that for all $R \geq R_0$, the set

$$S_R := \{x \in S \mid \|x\|^2 = R^2\}$$

is a non-empty compact set. Moreover, S_R is regular, i.e. for each $x \in S_R$, the vectors x and $\nabla f_i(x)$, $i \in I(x)$, are linearly independent.

Proof. The proof is similar to one of Lemma 2.1 and hence is omitted here. $\qquad\square$

4.2.3 *Frank–Wolfe type theorem*

The next result shows the existence of solutions to non-degenerate polynomial programs.

Theorem 4.3. *Let $F := (f_0, f_1, \dots, f_p)\colon \mathbb{R}^n \to \mathbb{R}^{p+1}$ be a polynomial map, which is convenient and non-degenerate at infinity. If f_0 is bounded from below on the non-empty set*

$$S := \{x \in \mathbb{R}^n \mid f_1(x) \leq 0, \dots, f_p(x) \leq 0\},$$

then f_0 attains its infimum on S.

The intuition behind the proof of Theorem 4.3 is as follows. By assumption, the restriction $f_0|_S$ is bounded from below, so it attains its infimum at some points in S or "at infinity". If $f_0|_S$ attains its infimum at infinity, it means that $f_0|_S$ has "singularities at infinity", so this possibility is eliminated by the condition of non-degeneracy at infinity (see Theorem 4.2).

Proof of Theorem 4.3. It suffices to show the following equality:

$$\lim_{r \to \infty} \min_{x \in S, \|x\|^2 = r^2} f_0(x) = +\infty.$$

By contradiction, suppose that there exists a sequence $\{x^k\}_{k\in\mathbb{N}} \subset S$ such that $\lim_{k\to\infty} \|x^k\| = \infty$, $\lim_{k\to\infty} f_0(x^k) = y \in \mathbb{R}$ and x^k is a solution of the following constrained polynomial optimization problem:

$$\min_{x\in S,\, \|x\|^2=r_k^2} f_0(x),$$

where $\{r_k\}_{k\in\mathbb{N}}$ is strictly increasing with $r_k \geq k$.

By Lemma 4.4, the set S is regular at infinity, so the set $S_k := S \cap \{\|x\|^2 = r_k^2\}, k \gg 1$, is regular, in view of Lemma 4.5. It follows from the KKT optimality conditions (Theorem 2.2) that there exist some real numbers $\lambda_i^k, i = 1, \ldots, p$, and μ_k such that

$$\nabla f_0(x^k) + \sum_{i=1}^{p} \lambda_i^k \nabla f_i(x^k) + \mu_k x^k = 0,$$

$$\lambda_i^k f_i(x^k) = 0, \quad \text{for } i = 1, \ldots, p.$$

Let

$$A := \big\{ (x, \lambda_0, \ldots, \lambda_p, \mu) \in \mathbb{R}^n \times \mathbb{R}^{p+2} \,|\, x \in S,\ \lambda_0 > 0,$$

$$\|(\lambda_0, \ldots, \lambda_p, \mu)\| = 1,$$

$$\lambda_0 \nabla f_0(x) + \sum_{i=1}^{p} \lambda_i \nabla f_i(x) + \mu x = 0,$$

$$\lambda_i f_i(x) = 0,\ \text{for } i = 1, \ldots, p \big\}.$$

Then A is a unbounded semi-algebraic set. By applying the Curve Selection Lemma at infinity (Theorem 1.12) for the semi-algebraic function $A \to \mathbb{R}, (x, \lambda_0, \ldots, \lambda_p, \mu) \mapsto f_0(x)$, we get a smooth semi-algebraic curve

$$(\varphi, \lambda_0, \ldots, \lambda_p, \mu) \colon (0, \epsilon)$$

$$\to \mathbb{R}^n \times \mathbb{R}^{p+2}, \quad t \mapsto (\varphi(t), \lambda_0(t), \ldots, \lambda_p(t), \mu(t)),$$

satisfying the following conditions:

(a) $\varphi(t) \in S, \lambda_0(t) > 0$ for $t \in (0, \epsilon)$, and $\|(\lambda_0(t), \ldots, \lambda_p(t), \mu(t))\| \equiv 1$;

(b) $\lim_{t \to 0^+} \|\varphi(t)\| = +\infty$ and $\lim_{t \to 0^+} f_0(\varphi(t)) = y$;

(c) $\lambda_0(t)\nabla f_0(\varphi(t)) + \sum_{i=1}^{p} \lambda_i(t)\nabla f_i(\varphi(t)) + \mu(t)\varphi(t) \equiv 0$; and

(d) $\lambda_i(t)f_i(\varphi(t)) \equiv 0$, for $i = 1, \ldots, p$.

Since the (smooth) functions λ_i and $f_i \circ \varphi$ are semi-algebraic, for $\epsilon > 0$ small enough, these functions are either constant or strictly monotone (see Theorem 1.8). Then, by condition (d), we can see that either $\lambda_i(t) \equiv 0$ or $f_i \circ \varphi(t) \equiv 0$; in particular,

$$\lambda_i(t)\frac{d}{dt}(f_i \circ \varphi)(t) \equiv 0, \quad i = 1, \ldots, p.$$

Replacing $\lambda_1(t), \ldots, \lambda_p(t)$, and $\mu(t)$ by $\lambda_1(t)/\lambda_0(t), \ldots, \lambda_p(t)/\lambda_0(t)$, and $\mu(t)/\lambda_0(t)$, respectively, we may assume that $\lambda_0(t) \equiv 1$. Let $I := \{i \in \{1, \ldots, p\} \mid f_i \circ \varphi(t) \equiv 0\}$. Then $\lambda_i(t) \equiv 0$ for $i \notin I$. It follows from condition (c) that

$$\frac{\mu(t)}{2}\frac{d\|\varphi(t)\|^2}{dt} = \mu(t)\left\langle \varphi(t), \frac{d\varphi}{dt} \right\rangle$$

$$= -\left\langle \nabla f_0(\varphi(t)), \frac{d\varphi}{dt} \right\rangle - \sum_{i \in I} \lambda_i(t)\left\langle \nabla f_i(\varphi(t)), \frac{d\varphi}{dt} \right\rangle$$

$$= -\frac{d}{dt}(f_0 \circ \varphi)(t) - \sum_{i \in I} \lambda_i(t)\frac{d}{dt}(f_i \circ \varphi)(t)$$

$$= -\frac{d}{dt}(f_0 \circ \varphi)(t).$$

Therefore, by condition (c) again,

$$\left| \frac{d}{dt}(f_0 \circ \varphi)(t) \right| = \left| \frac{\mu(t)}{2}\frac{d\|\varphi(t)\|^2}{dt} \right|$$

$$= \frac{\|\nabla f_0(\varphi(t)) + \sum_{i \in I} \lambda_i(t)\nabla f_i(\varphi(t))\|}{2\|\varphi(t)\|}\left| \frac{d\|\varphi(t)\|^2}{dt} \right|.$$

Note that the function $t \mapsto f_0 \circ \varphi(t)$ is not constant since otherwise we have $\mu(t) \equiv 0$, and by condition (c), the constrained set S is not regular at infinity, which is a contradiction. Now, by the Growth

Dichotomy Lemma (Lemma 1.7), we may write

$$\|\varphi(t)\| = c_1 t^\alpha + \text{higher order terms in } t,$$
$$f_0(\varphi(t)) = c_2 t^\beta + \text{higher order terms in } t,$$

for some non-zero constants c_1 and c_2. Condition (b) yields $\alpha < 0$ and $\beta \geq 0$. Moreover, we have

$$\left\| \nabla f_0(\varphi(t)) + \sum_{i \in I} \lambda_i(t) \nabla f_i(\varphi(t)) \right\|$$

$$= 2\|\varphi(t)\| \left| \frac{d}{dt}(f_0 \circ \varphi)(t) \right| \left| \frac{d\|\varphi(t)\|^2}{dt} \right|^{-1}$$

$$= ct^{\beta - \alpha} + \text{higher order terms in } t,$$

for some constant $c \neq 0$. Consequently, we get

$$\lim_{t \to 0^+} \left\| \nabla f_0(\varphi(t)) + \sum_{i \in I} \lambda_i(t) \nabla f_i(\varphi(t)) \right\| = 0.$$

Therefore, if we write $I = \{i_1, \ldots, i_q\} \subset \{1, \ldots, p\}$, then $(y, 0, \ldots, 0) \in \widetilde{K}_\infty(f_0, f_{i_1}, \ldots, f_{i_q})$, which contradicts Corollary 4.1. So we have proved that

$$\lim_{r \to \infty} \min_{x \in S, \|x\|^2 = r^2} f_0(x) = +\infty.$$

This implies immediately that f_0 is coercive on S, and so Theorem 4.3 follows. □

Remark 4.4. The assumption in Theorem 4.3 that the polynomial map $(f_0, f_1, \ldots, f_p)$ are convenient cannot be removed. A counterexample is $f_0(x, y) := x^2 + (xy - 1)^2$ and $S := \mathbb{R}^2$. It is easy to check that f_0 is non-degenerate at infinity. However, f_0 has 0 as unattainable infimum.

Finally, we conclude this section with an example illustrating Theorem 4.3.

Example 4.2. Let $n = 3$ and consider the polynomial

$$f_0(x, y, z) := x^8 + y^8 + z^8 + M(x, y, z),$$

where M is a polynomial of degree at most 7. Set $f_1(x, y, z) := 1 - ax^2 - by^2 - cz^2$, where a, b, c are non-zero real numbers and different from each other.

The Newton polyhedra at infinity of f_0 and f_1 are, respectively, the tetrahedra

$$\mathcal{N}(f_0) = \mathcal{N}\{(8, 0, 0), (0, 8, 0), (0, 0, 8)\},$$
$$\mathcal{N}(f_1) = \mathcal{N}\{(2, 0, 0), (0, 2, 0), (0, 0, 2)\}.$$

Hence, f_0 and f_1 are convenient. Let us check that the polynomial map $F := (f_0, f_1)$ is non-degenerate at infinity. The Minkowski sum

$$\mathcal{N}(F) = \mathcal{N}(f_0) + \mathcal{N}(f_1) = \mathcal{N}\{(10, 0, 0), (0, 10, 0), (0, 0, 10)\}$$

is again a tetrahedron. For simplicity, denote the convex hull of a set of points $a_1, \ldots, a_m \in \mathbb{R}^n$ by $[a_1, \ldots, a_m]$. Then $\mathcal{N}_\infty(F)$ has seven faces which are

$$\Delta^1 := [(10, 0, 0), (0, 10, 0), (0, 0, 10)]$$
$$= [(8, 0, 0), (0, 8, 0), (0, 0, 8)] + [(2, 0, 0), (0, 2, 0), (0, 0, 2)],$$
$$\Delta^2 := [(10, 0, 0), (0, 10, 0)] = [(8, 0, 0), (0, 8, 0)] + [(2, 0, 0), (0, 2, 0)],$$
$$\Delta^3 := [(10, 0, 0), (0, 0, 10)] = [(8, 0, 0), (0, 0, 8)] + [(2, 0, 0), (0, 0, 2)],$$
$$\Delta^4 := [(0, 10, 0), (0, 0, 10)] = [(0, 8, 0), (0, 0, 8)] + [(0, 2, 0), (0, 0, 2)],$$
$$\Delta^5 := [(10, 0, 0)] = [(8, 0, 0)] + [(2, 0, 0)],$$
$$\Delta^6 := [(0, 10, 0)] = [(0, 8, 0)] + [(0, 2, 0)],$$
$$\Delta^7 := [(0, 0, 10)] = [(0, 0, 8)] + [(0, 0, 2)].$$

It is clear that the following corresponding matrices all have rank 2 on $(\mathbb{R} \setminus \{0\})^3$:

$$A_{\Delta^1} := \begin{pmatrix} 8x^8 & 8y^8 & 8z^8 & x^8 + y^8 + z^8 & 0 \\ -2ax^2 & -2by^2 & -2cz^2 & 0 & -ax^2 - by^2 - cz^2 \end{pmatrix},$$

$$A_{\Delta^2} := \begin{pmatrix} 8x^8 & 8y^8 & 0 & x^8 + y^8 & 0 \\ -2ax^2 & -2by^2 & 0 & 0 & -ax^2 - by^2 \end{pmatrix},$$

$$A_{\Delta^3} := \begin{pmatrix} 8x^8 & 0 & 8z^8 & x^8 + z^8 & 0 \\ -2ax^2 & 0 & -2cz^2 & 0 & -ax^2 - cz^2 \end{pmatrix},$$

$$A_{\Delta^4} := \begin{pmatrix} 0 & 8y^8 & 8z^8 & y^8 + z^8 & 0 \\ 0 & -2by^2 & -2cz^2 & 0 & -by^2 - cz^2 \end{pmatrix},$$

$$A_{\Delta^5} := \begin{pmatrix} 8x^8 & 0 & 0 & x^8 & 0 \\ -2ax^2 & 0 & 0 & 0 & -ax^2 \end{pmatrix},$$

$$A_{\Delta^6} := \begin{pmatrix} 0 & 8y^8 & 0 & y^8 & 0 \\ 0 & -2by^2 & 0 & 0 & -by^2 \end{pmatrix},$$

$$A_{\Delta^7} := \begin{pmatrix} 0 & 0 & 8z^8 & z^8 & 0 \\ 0 & 0 & -2cz^2 & 0 & -cz^2 \end{pmatrix}.$$

Hence, the map F is non-degenerate at infinity. Furthermore, it is easy to check that f_0 is bounded from below on $\mathbb{R}^3$. By Theorem 4.3, f_0 attains its infimum on $\{(x, y, z) \in \mathbb{R}^3 \mid f_1(x, y, z) \leq 0\}$.

Bibliographic notes

The results in the first section are contained in a paper by Belousov and Klatte [10]. The material in the second section is taken from a paper by Đinh, Hà and Phạm [28].

In 1956, Frank and Wolfe [39] extended the fundamental existence theorem of linear programming by proving that an arbitrary quadratic function, bounded below on a non-empty polyhedral convex set, attains its infimum there. Many other authors (see, for example, [3, 8, 9, 95, 115, 125]) generalized the Frank–Wolfe theorem to broader classes of functions.

The partition problem, which is known to be NP-complete, can be found in [42]. The set $\widetilde{K}_\infty(F)$ of asymptotic critical values of F was defined in [131]. For more information concerning this set, we refer the reader to [49, 73].

Chapter 5

Well-Posedness in Unconstrained Polynomial Optimization

Abstract

We consider the class of unconstrained polynomial optimization problems, in which every problem of the class is obtained by perturbations of coefficients of the cost function. The main result states that there exists an open and dense semi-algebraic set $\mathcal{U}_N$ in the corresponding Euclidean space of data such that for every polynomial $f \in \mathcal{U}_N$, the problem of minimization of f over $\mathbb{R}^n$ is strongly well-posed.

5.1 Openness and density of Morse polynomials

Let $f\colon \mathbb{R}^n \to \mathbb{R}$ be a map of class C^∞. Recall that a point $x^0 \in \mathbb{R}^n$ is called a *critical point* of f if we have $x^0 \in \Sigma(f)$, i.e.

$$\frac{\partial f}{\partial x_1}(x^0) = \cdots = \frac{\partial f}{\partial x_n}(x^0) = 0.$$

A critical point $x^0 \in \mathbb{R}^n$ is called *non-degenerate* if the rank of the Hessian $\nabla^2 f(x^0)$ of f at x^0 is equal to n. The following notion plays an important role in this chapter.

Definition 5.1. The function f is said to be a *Morse function* if it has only non-degenerate critical points with distinct critical values.

It follows from this definition that the set $\Sigma(f)$ of critical points is discrete in $\mathbb{R}^n$ and the set $K_0(f)$ of critical values is discrete in $\mathbb{R}$. In particular, if f is polynomial, then these sets are finite.

In this section, we will show that the class of Morse polynomials forms an open, dense semi-algebraic subset of all polynomials with fixed Newton polyhedra.

For convenience, we repeat the definitions from Section 3.7. Recall that a subset $\mathcal{N} \subset \mathbb{R}^n_+$ is said to be a *Newton polyhedron at infinity* if there exists some finite subset $A \subset \mathbb{N}^n$ such that $\mathcal{N}$ is equal to the convex hull in $\mathbb{R}^n$ of $A \cup \{0\}$. A polyhedron $\mathcal{N} \subset \mathbb{R}^n_+$ is said to be *convenient* if it intersects each coordinate axis at a point different from the origin. Let $f \colon \mathbb{R}^n \to \mathbb{R}$ be a polynomial function. Suppose that f is written as $f = \sum_\alpha u_\alpha x^\alpha$ for some $u_\alpha \in \mathbb{R}$. The *Newton polyhedron at infinity* of f, denoted by $\mathcal{N}(f)$, is defined to be the convex hull in $\mathbb{R}^n$ of the set $\{\alpha \mid u_\alpha \neq 0\} \cup \{0\}$.

Let $\mathcal{N}$ be a convenient Newton polyhedron at infinity. We shall use the following notation:

$$\mathcal{A}_\mathcal{N} := \{f \in \mathbb{R}[x] \mid \mathcal{N}(f) \subseteq \mathcal{N}\},$$

$$L := \text{the set of points in } \mathcal{N} \text{ with integer coordinates},$$

$$N := \text{the cardinal number of the set } L.$$

Note that $N \geq n$ because the polyhedron $\mathcal{N}$ is convenient.

By using the lexicographic ordering on the set of monomials $x^\alpha, \alpha \in L$, for each $x \in \mathbb{R}^n$ we define the corresponding vector $\text{vec}(x) := (x^\alpha)_{\alpha \in L} \in \mathbb{R}^N$. For convenience, we identify each polynomial $f(x) = \sum_{\alpha \in L} u_\alpha x^\alpha \in \mathcal{A}_\mathcal{N}$ with its vector of coefficients $f := (u_\alpha)_{\alpha \in L} \in \mathbb{R}^N$, and so $f(x) = \langle f, \text{vec}(x) \rangle$. Then $\mathcal{A}_\mathcal{N}$ is identified with the Euclidean space $\mathbb{R}^N$.

Theorem 5.1. *There exists an open and dense semi-algebraic subset $\mathcal{B}_\mathcal{N}$ of $\mathcal{A}_\mathcal{N}$ such that for every $f \in \mathcal{B}_\mathcal{N}$, f is a Morse function.*

Proof. Consider the semi-algebraic map

$$\Phi \colon \mathcal{A}_\mathcal{N} \times \mathbb{R}^n \to \mathbb{R}^n, \quad (f, x) \longmapsto \nabla f(x),$$

where ∇f stands for the gradient of f.

Let $f := (u_\alpha)_{\alpha \in L} \in \mathcal{A}_\mathcal{N}$. Since $\mathcal{N}$ is convenient, $\alpha \in L$ for $|\alpha| = 1$. A simple calculation shows that

$$\left(\frac{\partial \Phi_i}{\partial u_\alpha}\right)_{|\alpha|=1,\, i=1,2,\ldots,n}$$

is the unit matrix of order n, and so rank $D\Phi(f,x) = n$ for all $(f,x) \in \mathcal{A}_\mathcal{N} \times \mathbb{R}^n$. In particular, $0 \in \mathbb{R}^n$ is a regular value of Φ. By Sard's theorem with parameter (Theorem 1.10), there exists a semi-algebraic set Σ_1 in $\mathcal{A}_\mathcal{N}$ of dimension at most $\dim \mathcal{A}_\mathcal{N} - 1$ such that for each $f \in \mathcal{A}_\mathcal{N} \backslash \Sigma_1$, 0 is a regular value of the map

$$\Phi_f \colon \mathbb{R}^n \to \mathbb{R}^n, \quad x \mapsto \nabla f(x).$$

This, together with Proposition 1.5, implies easily that the set $\widetilde{\mathcal{B}}_\mathcal{N} := \mathcal{A}_\mathcal{N} \backslash \overline{\Sigma}_1$ is an open and dense semi-algebraic subset of $\mathcal{A}_\mathcal{N}$. Furthermore, for all $f \in \widetilde{\mathcal{B}}_\mathcal{N}$, f has only non-degenerate critical points.

Next, consider the semi-algebraic map

$$\Psi \colon \widetilde{\mathcal{B}}_\mathcal{N} \times ((\mathbb{R}^n \times \mathbb{R}^n) \backslash \triangle) \to \mathbb{R} \times \mathbb{R}^n \times \mathbb{R}^n,$$
$$(f,x,y) \mapsto (f(x) - f(y), \nabla f(x), \nabla f(y)),$$

where we put

$$\triangle := \{(x,y) \in \mathbb{R}^n \times \mathbb{R}^n : x = y\}.$$

A direct computation shows that

$$D\Psi(f,x,y) = \begin{pmatrix} \text{vec}(x) - \text{vec}(y) & [\nabla f(x)]^T & -[\nabla f(y)]^T \\ \bullet & \nabla^2 f(x) & 0 \\ \bullet & 0 & \nabla^2 f(y) \end{pmatrix},$$

where $\nabla^2 f$ stands for the Hessian of f.

Let $f \in \widetilde{\mathcal{B}}_\mathcal{N}$ and $(x,y) \in (\mathbb{R}^n \times \mathbb{R}^n) \backslash \triangle$ be such that $\nabla f(x) = \nabla f(y) = 0$. Then

$$\text{vec}(x) - \text{vec}(y) \neq 0 \quad \text{and} \quad \text{rank}\,\nabla^2 f(x) = \text{rank}\,\nabla^2 f(y) = n,$$

and so rank $D\Psi(f,x,y) = 2n + 1$. Consequently, $0 \in \mathbb{R} \times \mathbb{R}^n \times \mathbb{R}^n$ is a regular value of Ψ. By Sard's theorem with parameter

(Theorem 1.10), there exists a semi-algebraic set Σ_2 in $\widetilde{\mathcal{B}}_N$ of dimension at most $\dim \widetilde{\mathcal{B}}_N - 1$ such that, for each $f \in \widetilde{\mathcal{B}}_N \backslash \Sigma_2$, 0 is a regular value of the map

$$\Psi_f \colon (\mathbb{R}^n \times \mathbb{R}^n) \backslash \triangle \to \mathbb{R} \times \mathbb{R}^n \times \mathbb{R}^n,$$
$$(x, y) \mapsto (f(x) - f(y), \nabla f(x), \nabla f(y)).$$

Let $\mathcal{B}_N := \widetilde{\mathcal{B}}_N \backslash \overline{\Sigma}_2$. In view of Proposition 1.5, $\mathcal{B}_N$ is an open and dense semi-algebraic subset of $\widetilde{\mathcal{B}}_N$. Note that

$$\dim(\mathbb{R} \times \mathbb{R}^n \times \mathbb{R}^n) = 2n + 1 > 2n = \dim((\mathbb{R}^n \times \mathbb{R}^n) \backslash \triangle).$$

Therefore, we have for all $f \in \mathcal{B}_N$, $\Psi_f^{-1}(0) = \emptyset$. Note that $\Psi_f^{-1}(0) = \emptyset$ if and only if the following system has no solution:

$$x \ne y, \quad f(x) - f(y) = 0, \quad \nabla f(x) = 0, \quad \nabla f(y) = 0.$$

This clearly completes the proof. $\qquad\qquad\qquad\qquad\qquad\qquad$ $\square$

5.2   Openness and density of non-degenerate polynomials

In this section, we show that the class of polynomials (with fixed Newton polyhedra), which are non-degenerate at infinity, is an open and dense semi-algebraic set in the corresponding Euclidean space of data.

Let $f \colon \mathbb{R}^n \to \mathbb{R}$ be a polynomial function. Recall that the *Newton boundary at infinity* of f, denoted by $\mathcal{N}_\infty(f)$, is defined as the set of the faces of the Newton polyhedron $\mathcal{N}(f)$ of f which do not contain the origin in $\mathbb{R}^n$. For each face $\triangle$ of $\mathcal{N}_\infty(f)$, we denote by $f_\triangle$ the polynomial $\sum_{\alpha \in \triangle} u_\alpha x^\alpha$.

Remark 5.1. By definition, if f is bounded from below on $\mathbb{R}^n$, then all vertices of $\mathcal{N}(f)$ have *even* non-negative integer coordinates; if f is coercive, then the polyhedron $\mathcal{N}(f)$ is convenient. We leave the (straightforward) proofs to the reader.

The following condition, which is a special case of non-degenerate conditions considered in Section 3.7, plays an important role in this chapter.

Definition 5.2. We say that f is *Kouchnirenko non-degenerate at infinity* if and only if

$$\left\{ x \in \mathbb{R}^n \left| \frac{\partial f_\Delta}{\partial x_1}(x) = \cdots = \frac{\partial f_\Delta}{\partial x_n}(x) = 0 \right. \right\} \subset \{ x \in \mathbb{R}^n \,|\, x_1 \cdots x_n = 0 \}$$

for all faces Δ of $\mathcal{N}_\infty(f)$.

Remark 5.2. By definition, it suffices to verify the non-degeneracy of f for faces of dimension at least 1.

Below are some examples illustrating this notion.

Example 5.1. (i) Let $n = 2$ and consider the dehomogenized Motzkin polynomial

$$f(x, y) := 1 + x^4 y^2 + x^2 y^4 - 3x^2 y^2.$$

Then the polyhedron $\mathcal{N}_\infty(f)$ is just the segment Δ connecting the points $(4, 2)$ and $(2, 4)$. Hence, it suffices to verify the non-degeneracy of f on Δ. By definition, $f_\Delta = x^4 y^2 + x^2 y^4$ and hence

$$\left\{ (x, y) \in \mathbb{R}^2 \left| \frac{\partial f_\Delta}{\partial x}(x) = \frac{\partial f_\Delta}{\partial y}(x) = 0 \right. \right\} = \{xy = 0\}.$$

Therefore, f is Kouchnirenko non-degenerate at infinity.

(ii) Let $n = 3$ and consider the Robinson polynomial

$$f(x, y, z) := x^6 + y^6 + z^6 + 3x^2 y^2 z^2$$
$$- x^4(y^2 + z^2) - y^4(z^2 + x^2) - z^4(x^2 + y^2).$$

By definition, $\mathcal{N}(f)$ is the convex hull of the points $(0, 0, 0)$, $(6, 0, 0), (0, 6, 0)$, and $(0, 0, 6)$. Let Δ be the segment connecting the two points $(6, 0, 0)$ and $(0, 6, 0)$. Then Δ is a face of $\mathcal{N}_\infty(f)$ and the corresponding polynomial $f_\Delta(x, y, z) = x^6 + y^6 - x^4 y^2 - y^4 x^2$. It is not hard to verify that

$$\left\{ (x, y, z) \in \mathbb{R}^3 \left| \frac{\partial f_\Delta}{\partial x} = \frac{\partial f_\Delta}{\partial y} = \frac{\partial f_\Delta}{\partial y} = 0 \right. \right\} \supset \{x = y\}.$$

Hence, the polynomial f is not Kouchnirenko non-degenerate at infinity.

(iii) Let $f(x) := \sum_{j=1}^{n} x_j^d + g(x)$, where g is a polynomial of degree at most $d - 1$. By definition, $\mathcal{N}(f)$ is the convex hull of the origin and the points

$$(d, 0, \ldots, 0), (0, d, \ldots, 0), \ldots, (0, 0, \ldots, d).$$

In particular, $\mathcal{N}_\infty(f)$ is the convex hull of the latter n points.

Take an arbitrary face Δ of $\mathcal{N}_\infty(f)$. It is easy to see that $f_\Delta(x) = \sum_{j \in J} x_j^d$ for some non-empty subset $J \subset \{1, \ldots, n\}$, which yields

$$\left\{ x \in \mathbb{R}^n \,\middle|\, \frac{\partial f_\Delta}{\partial x_1}(x) = \cdots = \frac{\partial f_\Delta}{\partial x_n}(x) = 0 \right\}$$

$$= \{ x \in \mathbb{R}^n \mid x_j = 0 \text{ for } j \in J \}.$$

Therefore, as Δ is arbitrary, f is Kouchnirenko non-degenerate at infinity.

Let $\mathcal{N}$ be a Newton polyhedron at infinity and define

$$\mathcal{A}_\mathcal{N} := \{ f \in \mathbb{R}[x] \mid \mathcal{N}(f) \subseteq \mathcal{N} \}.$$

As in the previous section, we identify each polynomial $f(x) = \sum_{\alpha \in L} u_\alpha x^\alpha \in \mathcal{A}_\mathcal{N}$ with its vector of coefficients $f := (u_\alpha)_{\alpha \in L} \in \mathbb{R}^N$, where L is the set of points in $\mathcal{N}$ with integer coordinates and $N := \#L$. In particular, for each face $\Delta \in \mathcal{N}_\infty$, the principal part $f_\Delta := \sum_{\alpha \in \Delta} u_\alpha x^\alpha$ of f with respect to Δ is identified with the vector $(u_\alpha)_{\alpha \in \Delta}$, and we write $\mathbb{R}^\Delta$ for the set of vectors $f_\Delta := (u_\alpha)_{\alpha \in \Delta}$. The following result shows that polynomials, which are Kouchnirenko non-degenerate at infinity, are generic.

Theorem 5.2. *With the above notations, we have*

$$\mathcal{C}_\mathcal{N} := \{ f \in \mathcal{A}_\mathcal{N} \mid \mathcal{N}(f) = \mathcal{N} \text{ and }$$

$$f \text{ is Kouchnirenko non-degenerate at infinity} \}$$

is an open and dense semi-algebraic subset of $\mathcal{A}_\mathcal{N}$.

Proof. For each face $\Delta \in \mathcal{N}_\infty$, let Φ_Δ denote the semi-algebraic function

$$\Phi_\Delta \colon \mathbb{R}^\Delta \times (\mathbb{R}\setminus\{0\})^n \to \mathbb{R}, \quad (f_\Delta, x) \mapsto f_\Delta(x).$$

It is easy to check that $0 \in \mathbb{R}$ is a regular value of Φ_Δ. By Sard's theorem with parameter 1.10, there exists a semi-algebraic set Σ_Δ in $\mathbb{R}^\Delta$ of dimension at most $\dim \mathbb{R}^\Delta - 1$ such that for each $f_\Delta \in \mathbb{R}^\Delta \backslash \Sigma_\Delta$, 0 is a regular value of the function

$$\Phi_{\Delta, f_\Delta} \colon (\mathbb{R} \backslash \{0\})^n \to \mathbb{R}, \quad x \mapsto f_\Delta(x).$$

A simple calculation shows that this function is quasi-homogeneous of non-zero degree. It follows from the theorem of Euler on homogeneous functions that for each $f_\Delta \in \mathbb{R}^\Delta \backslash \Sigma_\Delta$, we have

$$\nabla \Phi_{\Delta, f_\Delta}(x) \neq 0 \quad \text{for all } x \in (\mathbb{R} \backslash \{0\})^n.$$

Now, if $\mathcal{A}_\mathcal{N} \to \mathbb{R}^\Delta, f \mapsto f_\Delta$, denotes the projection onto $\mathbb{R}^\Delta$,

$$\mathcal{U}_\Delta := \{f \in \mathcal{A}_\mathcal{N} \, | \, f_\Delta \in \mathbb{R}^\Delta \backslash \overline{\Sigma}_\Delta\}$$

is an open and dense semi-algebraic subset of $\mathcal{A}_\mathcal{N}$. By definition, $\bigcap_{\Delta \in \mathcal{N}_\infty} \mathcal{U}_\Delta \subset \mathcal{C}_\mathcal{N}$. Therefore, the set $\mathcal{C}_\mathcal{N}$ is dense in $\mathcal{A}_\mathcal{N}$.

Next, we show that $\mathcal{C}_\mathcal{N}$ is an open semi-algebraic set. To do this, for every face $\Delta \in \mathcal{N}_\infty$, we define

$$\mathcal{X}(\Delta) := \left\{(f, x) \in \mathcal{A}_\mathcal{N} \times \mathbb{R}^n \, \middle| \, \frac{\partial f_\Delta}{\partial x_1}(x) = \cdots = \frac{\partial f_\Delta}{\partial x_n}(x) = 0 \right\},$$

$$\mathcal{X}(\Delta)^* := \mathcal{X}(\Delta) \cap \{(f, x) \in \mathcal{A}_\mathcal{N} \times \mathbb{R}^n \, | \, x_1 \cdots x_n \neq 0\}.$$

Note that $\mathcal{X}(\Delta)$ is closed and that $\overline{\mathcal{X}(\Delta)^*} = \mathcal{X}(\Delta)$. Let us consider the union $\mathcal{X}^* := \bigcup_{\Delta \in \mathcal{N}_\infty} \mathcal{X}(\Delta)^*$ and the projection $\pi \colon \mathcal{A}_\mathcal{N} \times \mathbb{R}^n \to \mathcal{A}_\mathcal{N}, (f, x) \mapsto f$. One observes that $\Sigma := \pi(\mathcal{X}^*)$ is a semi-algebraic set, since it is the projection of a semi-algebraic set (Theorem 1.5). Furthermore, we have $\mathcal{C}_\mathcal{N} = \widetilde{\mathcal{A}}_\mathcal{N} \backslash \Sigma$, where we put

$$\widetilde{\mathcal{A}}_\mathcal{N} := \left\{f = \sum_{\alpha \in L} u_\alpha x^\alpha \in \mathcal{A}_\mathcal{N} \, \middle| \, u_\alpha \neq 0 \text{ for all } \alpha \in V \right\}$$

with V being the set of vertices of $\mathcal{N}$. Therefore, $\mathcal{C}_\mathcal{N}$ is semi-algebraic because the set $\widetilde{\mathcal{A}}_\mathcal{N}$ is (open and dense) semi-algebraic in $\mathcal{A}_\mathcal{N}$.

To complete the proof of the theorem, we need to show that $\mathcal{C}_\mathcal{N}$ is an open set, or equivalently, Σ is a closed set. To see this, let $f^0 := (u_\alpha^0)_{\alpha \in L} \in \overline{\Sigma}$. By the Curve Selection Lemma at infinity (Theorem 1.12), there exists a face Δ of $\mathcal{N}_\infty$ and a smooth semi-algebraic curve $((u_\alpha(t))_{\alpha \in L}, \varphi(t)) \in \mathcal{X}(\Delta)^*$ defined on a small enough interval $(0, \epsilon)$ such that

(a) $\lim_{t \to 0^+} u_\alpha(t) = u_\alpha^0$ for all $\alpha \in L$; and
(b) either $\lim_{t \to 0^+} \|\varphi(t)\| = +\infty$ or $\lim_{t \to 0^+} \varphi(t) = a \in \mathcal{X}(\Delta)$.

By the Growth Dichotomy Lemma (Lemma 1.7), for each $j \in \{1, \dots, n\}$, we can expand the coordinate φ_j as follows:

$$\varphi_j(t) = x_j^0 t^{q_j} + \text{higher order terms in } t,$$

where $x_j^0 \neq 0$ and $q_j \in \mathbb{Z}$. (Note that $\varphi_j \not\equiv 0$ for all j.)

Let d be the minimal value of the linear function $\sum_{j=1}^n q_j \kappa_j$ on Δ, and let $\widetilde{\Delta}$ be the (unique) maximal face of Δ where the linear function takes this value. Clearly, $\widetilde{\Delta}$ is a face of $\mathcal{N}_\infty$. For each $t \in (0, \epsilon)$, we define $f_t \in \mathbb{R}[x]$ to be $(u_\alpha(t))_{\alpha \in L}$, i.e. $f_t(x) = \sum_{\alpha \in L} u_\alpha(t) x^\alpha$. By a direct calculation, then

$$0 \equiv \frac{\partial f_{t,\Delta}}{\partial x_j}(\varphi(t)) = \frac{\partial f_{t,\widetilde{\Delta}}}{\partial x_j}(x^0) t^{d-q_j} + \text{higher order terms in } t,$$

where $x^0 := (x_1^0, \dots, x_n^0) \in (\mathbb{R} \backslash 0)^n$. By taking the limit $f_t \to f^0$ and focusing on the first terms of the above expansions, we get that $(f^0, x^0) \in \mathcal{X}(\widetilde{\Delta})^* \subset \mathcal{X}^*$. Thus $f^0 \in \Sigma$, which concludes the proof that $\Sigma = \overline{\Sigma}$. $\qquad\square$

5.3 Non-degeneracy and coercivity

Let $\mathcal{N}$ be a convenient Newton polyhedron at infinity. We shall use the notations:

$$\mathcal{A}_\mathcal{N} := \{ f \in \mathbb{R}[x] \,|\, \mathcal{N}(f) \subseteq \mathcal{N} \},$$

$$V := \text{the set of vertices of } \mathcal{N},$$

$$L := \text{the set of points in } \mathcal{N} \text{ with integer coordinates},$$

$$N := \text{the cardinal number of the set } L.$$

As shown in the previous section,

$$\mathcal{C}_{\mathcal{N}} := \{f \in \mathcal{A}_{\mathcal{N}} \,|\, \mathcal{N}(f) = \mathcal{N} \text{ and}$$

$$f \text{ is Kouchnirenko non-degenerate at infinity}\}$$

is an open and dense semi-algebraic subset of $\mathcal{A}_{\mathcal{N}}$.

Lemma 5.1. *Let $f \in \mathcal{C}_{\mathcal{N}}$ and suppose that f is bounded from below on $\mathbb{R}^n$. Then, for each face $\Delta \in \mathcal{N}_{\infty}(f)$, $f_{\Delta} \geq 0$ on $\mathbb{R}^n$ and $f_{\Delta} > 0$ on $(\mathbb{R}\backslash 0)^n$.*

Proof. Let Δ be an arbitrary face of $\mathcal{N}_{\infty}(f)$. We first show that $f_{\Delta} \geq 0$ on $\mathbb{R}^n$. In fact, since f_{Δ} is continuous, it suffices to prove that $f_{\Delta} \geq 0$ on $(\mathbb{R}\backslash 0)^n$.

By contradiction, suppose that there exists a point $x^0 \in (\mathbb{R}\backslash 0)^n$ such that $f_{\Delta}(x^0) < 0$. Let J be the smallest subset of $\{1, \ldots, n\}$ such that $\Delta \in \mathcal{N}_{\infty}(f) \cap \mathbb{R}^J$. Then, there exists a non-zero vector $q \in \mathbb{R}^n$, with $\min_{j \in J} q_j < 0$, such that

$$\Delta = \left\{\alpha \in \mathcal{N}_{\infty}(f) \cap \mathbb{R}^J \,\middle|\, \langle q, \alpha \rangle = \min_{\beta \in \mathcal{N}_{\infty}(f) \cap \mathbb{R}^J} \langle q, \beta \rangle\right\}.$$

Let ϵ be a positive real number small enough and define the monomial curve $\varphi(t) \colon (0, \varepsilon) \to \mathbb{R}^n, t \mapsto (\varphi_1(t), \ldots, \varphi_n(t))$, by

$$\varphi_j(t) = \begin{cases} x_j^0 t^{q_j} & \text{if } j \in J, \\ 0 & \text{otherwise.} \end{cases}$$

Let $d := \min_{\alpha \in \mathcal{N}_{\infty}(f) \cap \mathbb{R}^J} \langle q, \alpha \rangle$. It is clear that $d < 0$. Moreover, we can write

$$f(\varphi(t)) = f_{\Delta}(x^0) t^d + \text{higher order terms in } t.$$

Since $f_{\Delta}(x^0) < 0$, it holds that

$$\lim_{t \to 0^+} f(\varphi(t)) = -\infty,$$

which is a contradiction.

Next, we show that $f_{\Delta} > 0$ on $(\mathbb{R}\backslash 0)^n$. Indeed, it was proved that $f_{\Delta} \geq 0$ on $\mathbb{R}^n$. So, by contradiction, suppose that there exists a point $x^0 \in (\mathbb{R}\backslash\{0\})^n$ such that $f_{\Delta}(x^0) = 0$. It follows that x^0 is a global minimizer of f_Δ on $\mathbb{R}^n$. Consequently, x_0 is a critical point of f_Δ, i.e. $\frac{\partial f_\Delta}{\partial x_j}(x^0) = 0$ for $j = 1, \ldots, n$, which contradicts the non-degeneracy of f. □

We define the semi-algebraic function $P_\mathcal{N} \colon \mathbb{R}^n \to \mathbb{R}$ by setting

$$P_\mathcal{N}(x) := \sum_{\alpha \in V} |x^\alpha|.$$

Lemma 5.2. *For all* $x \in \mathbb{R}^n$, *it holds that*

$$P_\mathcal{N}(x) \leq \sum_{\alpha \in L} |x^\alpha| \leq N P_\mathcal{N}(x).$$

In particular, $P_\mathcal{N}$ *is coercive on* $\mathbb{R}^n$.

Proof. Since $V \subset L$, we have for all $x \in \mathbb{R}^n$,

$$P_\mathcal{N}(x) = \sum_{\alpha \in V} |x^\alpha| \leq \sum_{\alpha \in L} |x^\alpha|,$$

which proves the first inequality.

Let $v^1, \ldots, v^k$ be the vertices of the polyhedron $\mathcal{N}$, i.e. $V = \{v^1, \ldots, v^k\}$. Take any $\alpha \in L$. Then there exist non-negative real numbers $\lambda_1, \ldots, \lambda_k$, with $\lambda_1 + \cdots + \lambda_k = 1$, such that

$$\alpha = \lambda_1 v^1 + \cdots + \lambda_k v^k.$$

We have for all $x \in \mathbb{R}^n$,

$$|x^\alpha| = |x^{\lambda_1 v^1 + \cdots + \lambda_k v^k}| = (|x^{v^1}|)^{\lambda_1} \cdots (|x^{v^k}|)^{\lambda_k}$$

$$\leq \lambda_1 |x^{v^1}| + \cdots + \lambda_k |x^{v^k}|$$

$$\leq |x^{v^1}| + \cdots + |x^{v^k}| = P_\mathcal{N}(x).$$

Hence, $\sum_{\alpha \in L} |x^\alpha| \leq N P_\mathcal{N}(x)$, which proves the second inequality. This, together with the fact that the polyhedron $\mathcal{N}$ is convenient, implies that $\lim_{\|x\| \to +\infty} P_\mathcal{N}(x) = +\infty$. □

A relation between the non-degeneracy and coercivity conditions is given in the following theorem.

Theorem 5.3. *Let $f \in C_N$ and suppose that f is bounded from below on $\mathbb{R}^n$. Then there exist positive constants c_1, c_2, and r such that*

$$c_1 P_N(x) \leq f(x) \leq c_2 P_N(x) \quad \text{for all } x \in \mathbb{R}^n, \; \|x\| \geq r.$$

In particular, f is coercive on $\mathbb{R}^n$.

Proof. Let us write $f(x) = \sum_{\alpha \in L} u_\alpha x^\alpha$. Then

$$f(x) \leq |f(x)| \leq \sum_{\alpha \in L} |u_\alpha x^\alpha| \leq c \sum_{\alpha \in L} |x^\alpha|,$$

where $c := \max_{\alpha \in L} |u_\alpha|$. It follows from Lemma 5.2 that $f(x) \leq cN P_N(x)$ for all $x \in \mathbb{R}^n$, which yields the second inequality.

We now prove the first inequality. By contradiction and using the Curve Selection Lemma at infinity (Theorem 1.12), we get an analytic curve $\varphi \colon (0, \varepsilon) \to \mathbb{R}^n, t \mapsto (\varphi_1(t), \ldots, \varphi_n(t))$, with $\lim_{t \to 0^+} \|\varphi(t)\| = +\infty$, such that

$$\lim_{t \to 0^+} \frac{f(\varphi(t))}{P_N(\varphi(t))} = 0. \tag{5.1}$$

Let $J := \{j \mid \varphi_j \not\equiv 0\} \subset \{1, \ldots, n\}$. For $j \in J$, we can expand the coordinate function φ_j in terms of the parameter, say:

$$\varphi_j(t) = x_j^0 t^{q_j} + \text{higher order terms in } t,$$

where $x_j^0 \neq 0$ and $q_j \in \mathbb{Q}$. We have $\min_{j \in J} q_j < 0$, since $\lim_{t \to 0^+} \|\varphi(t)\| = +\infty$.

Recall that $\mathbb{R}^J := \{\alpha = (\alpha_1, \ldots, \alpha_n) \in \mathbb{R}^n \mid \alpha_j = 0 \text{ for } j \notin J\}$. Since the polyhedron $N(f) = N$ is convenient, $N_\infty(f) \cap \mathbb{R}^J \neq \emptyset$. Let d be the minimal value of the linear function $\sum_{j \in J} q_j \alpha_j$ on $N_\infty(f) \cap \mathbb{R}^J$, i.e. $d := \min_{\alpha \in N_\infty(f) \cap \mathbb{R}^J} \sum_{j \in J} q_j \alpha_j$ and let Δ be the (unique) maximal face of $N_\infty(f)$ where the linear function $\sum_{j \in J} a_j \alpha_j$ takes its minimal value, i.e.

$$\Delta = \left\{ \alpha \in N_\infty(f) \cap \mathbb{R}^J \;\middle|\; \sum_{j \in J} q_j \alpha_j = d \right\}.$$

By definition, then $d < 0$ and $\Delta \in N_\infty(f)$. We can write

$$f(\varphi(t)) = f_\Delta(x^0) t^d + \text{higher order terms in } t,$$

where $x^0 := (x_1^0, \ldots, x_n^0)$ and $x_j^0 := 1$ for $j \notin J$. Since $f \in \mathcal{C}_{\mathcal{N}}$ and f is bounded from below on $\mathbb{R}^n$, it follows from Lemma 5.1 that $f_\Delta(x^0) > 0$, and hence,

$$f(\varphi(t)) \simeq t^d \quad \text{as } t \to 0^+. \tag{5.2}$$

On the other hand, it is not hard to check that

$$P_{\mathcal{N}}(\varphi(t)) = \sum_{\alpha \in V} |\varphi(t)^\alpha|$$

$$= \left(\sum_{\alpha \in V \cap \Delta} |(x^0)^\alpha| \right) t^d + \text{higher order terms in } t.$$

Since $\sum_{\alpha \in V \cap \Delta} |(x^0)^\alpha| > 0$, we get that

$$P_{\mathcal{N}}(\varphi(t)) \simeq t^d \quad \text{as } t \to 0^+.$$

This, combined with (5.1) and (5.2), gives a contradiction. $\qquad\square$

5.4 Well-posedness in unconstrained polynomial optimization

This section deals with well-posedness in unconstrained polynomial optimization. In order to do this, let $\mathcal{N}$ be a convenient Newton polyhedron at infinity. As in the previous sections, we use the following notations:

$$\mathcal{A}_{\mathcal{N}} := \{ f \in \mathbb{R}[x] \, | \, \mathcal{N}(f) \subseteq \mathcal{N} \},$$

$$V := \text{the set of vertices of } \mathcal{N},$$

$$L := \text{the set of points in } \mathcal{N} \text{ with integer coordinates,}$$

$$N := \text{the cardinal number of the set } L.$$

We also define the semi-algebraic function $P_{\mathcal{N}} \colon \mathbb{R}^n \to \mathbb{R}$ by

$$P_{\mathcal{N}}(x) := \sum_{\alpha \in V} |x^\alpha|.$$

Now, we are ready to establish the main result of the chapter.

Theorem 5.4. *There exists an open and dense semi-algebraic subset* $\mathcal{U}_{\mathcal{N}}$ *of* $\mathcal{A}_{\mathcal{N}}$ ($\equiv \mathbb{R}^N$) *such that for all* $f \in \mathcal{U}_{\mathcal{N}}$, *which are bounded from* *below on* $\mathbb{R}^n$, *the problem* $\inf_{x \in \mathbb{R}^n} f(x)$ *is strongly well-posed in the sense that there exists* $\epsilon > 0$ *such that for all* $u := (u_\alpha) \in \mathbb{R}^N$, *with* $\|u\| < \epsilon$, *the following statements hold:*

(i) *There exist positive constants* c_1, c_2, *and* r *such that*

$$c_1 P_N(x) \leq f_u(x) \leq c_2 P_N(x) \quad \text{for all } x \in \mathbb{R}^n, \ \|x\| \geq r,$$

   where we put $f_u := f + \sum_{\alpha \in L} u_\alpha x^\alpha \in \mathcal{U}_N$.

(ii) f_u *is coercive.*

(iii) f_u *has a unique global minimizer* $x_u^* \in \mathbb{R}^n$.

(iv) f_u *is a Morse function, i.e. it has only non-degenerate critical points with distinct critical values.*

(v) *The corresponding* $\{u \in \mathbb{R}^N \mid \|u\| < \epsilon\} \to \mathbb{R}^n, u \mapsto x_u^*$, *is an analytic map with* $\lim_{u \to 0} x_u^* = x^*$, *where* x^* *is the unique global minimizer of* f *on* $\mathbb{R}^n$.

(vi) *For all* $x_u \in \mathbb{R}^n$, *if* $\lim_{u \to 0}[f_u(x_u) - \inf_{x \in \mathbb{R}^n} f_u(x)] = 0$ *then* $\lim_{u \to 0} x_u = x^*$.

Proof. Let $\mathcal{U}_N := \mathcal{B}_N \cap \mathcal{C}_N$, where $\mathcal{B}_N$ and $\mathcal{C}_N$ are the sets defined in Theorems 5.1 and 5.2. We have $\mathcal{U}_N$ which is an open and dense semi-algebraic subset of $\mathcal{A}_N$.

Take arbitrary $f \in \mathcal{U}_N$ and assume that f is bounded from below on $\mathbb{R}^n$, i.e. $\inf_{x \in \mathbb{R}^n} f(x) > -\infty$.

(i) By Theorem 5.3, there exist positive constants c_1', c_2', and r such that

$$c_1' P_N(x) \leq f(x) \leq c_2' P_N(x) \quad \text{for all } x \in \mathbb{R}^n, \ \|x\| \geq r. \quad (5.3)$$

Since the set $\mathcal{U}_N$ is open, there exists $\epsilon \in (0, c_1'/N)$ such that for all $u := (u_\alpha)_{\alpha \in L} \in \mathbb{R}^N$, with $\|u\| < \epsilon$, it holds that

$$f_u := f + \sum_{\alpha \in L} u_\alpha x^\alpha \in \mathcal{U}_N.$$

In particular, we have for all $x \in \mathbb{R}^n$,

$$f(x) - \sum_{\alpha \in L} |u_\alpha x^\alpha| \leq f_u(x) := f(x) + \sum_{\alpha \in L} u_\alpha x^\alpha$$

$$\leq f(x) + \sum_{\alpha \in L} |u_\alpha x^\alpha|.$$

Hence,

$$f(x) - \epsilon \sum_{\alpha \in L} |x^\alpha| \le f_u(x) \le f(x) + \epsilon \sum_{\alpha \in L} |x^\alpha|.$$

It follows from inequality (5.3) and Lemma 5.2 that

$$(c_1' - \epsilon N) P_{\mathcal{N}}(x)$$
$$\le f_u(x) \le (c_2' + \epsilon N) P_{\mathcal{N}}(x) \quad \text{for all } x \in \mathbb{R}^n, \ \|x\| \ge r.$$

Clearly, the positive constants $c_1 := c_1' - \epsilon N$ and $c_2 := c_2' + \epsilon N$ satisfy the required inequalities.

(ii) This is an immediate consequence of (i) and Lemma 5.2.

(iii) Since f_u is coercive, it has a global minimizer. This, combined with Theorem 5.1, implies the statement.

(iv) This is a direct consequence of Theorem 5.1.

(v) We first show that $\lim_{u \to 0} x_u^* = x^*$. Indeed, we see that if $\|x_u^*\| \ge r$, then

$$c_1 P_{\mathcal{N}}(x_u^*) \le f_u(x_u^*) \le f_u(x^*) = f(x^*) + \sum_{\alpha \in L} u_\alpha (x^*)^\alpha$$

$$\le f(x^*) + \epsilon \sum_{\alpha \in L} |(x^*)^\alpha|.$$

This, together with the coercivity of the polynomial $P_{\mathcal{N}}$ (see Lemma 5.2), implies that the set $\{x_u^* \mid \|u\| < \epsilon\}$ is bounded. Note that

$$f(x^*) + \sum_{\alpha \in L} u_\alpha (x_u^*)^\alpha \le f(x_u^*) + \sum_{\alpha \in L} u_\alpha (x_u^*)^\alpha = f_u(x_u^*)$$

$$\le f_u(x^*) = f(x^*) + \sum_{\alpha \in L} u_\alpha (x^*)^\alpha.$$

Consequently, we have $f(x^*) = \lim_{u \to 0} f_u(x_u^*)$. Finally, let y^* be an accumulation point of the sequence $\{x_u^*\}$ as $u \to 0$. Then

$$f(x^*) = \lim_{u \to 0} f_u(x_u^*) = \lim_{u \to 0} \left[f(x_u^*) + \sum_{\alpha \in L} u_\alpha (x_u^*)^\alpha \right]$$

$$= \lim_{u \to 0} f(x_u^*) = f(y^*),$$

which yields $y^* = x^*$ because x^* is the unique minimizer of f. Therefore, $\lim_{u \to 0} x_u^* = x^*$.

We next define the polynomial map $\Phi \colon \mathcal{U}_N \times \mathbb{R}^n \to \mathbb{R}^n, (u, x) \mapsto \Phi(u, x) := \nabla f_u(x)$, and consider the system of equations $\Phi(u, x) = 0$. We have

$$\Phi(0, x^*) = \nabla f(x^*) = 0 \quad \text{and} \quad \operatorname{rank} D\Phi(0, x^*) = \operatorname{rank} \nabla^2 f(x^*) = n.$$

By the implicit function theorem and by shrinking ϵ if necessary, there exists a unique analytic map

$$s \colon \{u \in \mathbb{R}^N \mid \|u\| < \epsilon\} \to \mathbb{R}^n, \quad u \mapsto s(u),$$

such that $s(0) = x^*$ and $\Phi(u, s(u)) = 0$ for $\|u\| < \epsilon$. On the other hand, since x_u^* is the global minimizer of f_u, $\Phi(x_u^*, u) = \nabla f_u(x_u^*) = 0$ for all $\|u\| < \epsilon$. By the uniqueness of the solution s of the system $\Phi(x, u) = 0$ in some neighborhood of the point $(x^*, 0)$, we get $s(u) = x_u^*$ for all $\|u\| < \epsilon$.

(vi) Let $\{x_u\} \subset \mathbb{R}^n$ be an arbitrary sequence with

$$\lim_{u \to 0} \left[f_u(x_u) - \inf_{x \in \mathbb{R}^n} f_u(x) \right] = 0.$$

We will show that $\lim_{u \to 0} x_u = x^*$. In fact, we have

$$|f_u(x_u) - f(x^*)| \leq |f_u(x_u) - f_u(x_u^*)| + |f_u(x_u^*) - f(x^*)|$$

$$= \left| f_u(x_u) - \inf_{x \in \mathbb{R}^n} f_u(x) \right| + |f_u(x_u^*) - f(x^*)|.$$

Consequently, $\lim_{u \to 0} f_u(x_u) = f(x^*)$. It follows from (i) that, if $\|x_u\| \geq r$, then $c_1 P_N(x_u) \leq f_u(x_u)$, and so $c_1 P_N(x_u) \leq f(x^*) + M$ for some sufficient large real number M. Note that $P_N(x)$ is coercive due to Lemma 5.2. Therefore, the set $\{x_u \mid \|u\| < \epsilon\}$ is bounded. Finally, let y^* be an accumulation point of the sequence $\{x_u\}$ as $u \to 0$. Then

$$f(x^*) = \lim_{u \to 0} f_u(x_u)$$

$$= \lim_{u \to 0} \left[f(x_u) + \sum_{\alpha \in L} u_\alpha(x_u)^\alpha \right] = \lim_{u \to 0} f(x_u) = f(y^*),$$

which yields $y^* = x^*$ because x^* is the unique minimizer of f. Therefore, $\lim_{u \to 0} x_u = x^*$. This completes the proof of the theorem. □

In the rest of this chapter, we would establish local and global sharp minima with explicit exponents for a large class of unconstrained polynomial optimization problems.

Note that the assumption that the polyhedron $\mathcal{N}$ is convenient implies the existence of positive integers m_i for $i = 1, \ldots, n$, such that all points $(0, \ldots, 0, m_i, 0, \ldots, 0) \in \mathbb{R}^n$ are vertices of $\mathcal{N}$. Let

$$m_{\mathcal{N}} := \min_{i=1,\ldots,n} m_i. \tag{5.4}$$

Corollary 5.1. *Let $\mathcal{U}_{\mathcal{N}}$ be the open and dense semi-algebraic set defined in Theorem 5.4. Then for any $f \in \mathcal{U}_{\mathcal{N}}$, if f is bounded from below on $\mathbb{R}^n$, then there exist positive constants δ, c_1 and c_2 such that the following inequalities hold:*

$$|f(x) - f(x^*)|^{\frac{1}{2}} \geq c_1 \|x - x^*\| \quad \text{for } \|x - x^*\| \leq \delta,$$

$$|f(x) - f(x^*)|^{\frac{1}{m_{\mathcal{N}}}} + |f(x) - f(x^*)|^{\frac{1}{2}} \geq c_2 \|x - x^*\| \quad \text{for } x \in \mathbb{R}^n,$$

where x^* is the unique global minimizer of f on $\mathbb{R}^n$ and $m_{\mathcal{N}}$ is defined in (5.4).

Proof. We have the Taylor expansion of f at x^*:

$$f(x) - f(x^*) = \langle \nabla^2 f(x^*)(x - x^*), x - x^* \rangle + o(\|x - x^*\|^2). \tag{5.5}$$

Since $f \in \mathcal{U}_{\mathcal{N}}$ and x^* is the unique global minimizer of f on $\mathbb{R}^n$, the Hessian $\nabla^2 f(x^*)$ of f at x^* is positive definite, and so there exists a constant $c_1' > 0$ such that

$$\langle \nabla^2 f(x^*)(x - x^*), x - x^* \rangle \geq c_1' \|x - x^*\|^2 \quad \text{for } x \in \mathbb{R}^n.$$

We deduce easily from inequality (5.5) that

$$|f(x) - f(x^*)|^{\frac{1}{2}} \geq c_1 \|x - x^*\| \quad \text{for } \|x - x^*\| \leq \delta \tag{5.6}$$

for sufficiently small real numbers $c_1 > 0$ and $0 < \delta < 1$, which proves the first inequality.

On the other hand, it follows from Theorem 5.3 that there exist constants $c_2' > 0$ and $r \geq 1$ such that

$$f(x) \geq c_2' P_{\mathcal{N}}(x) = c_2' \sum_{\alpha \in V} |x^\alpha| \geq c_2' \sum_{i=1}^{n} |x_i^{m_i}| \quad \text{for } \|x\| \geq r.$$

Since f is bounded from below on $\mathbb{R}^n$ and the polyhedron $\mathcal{N}(f) = \mathcal{N}$ is convenient, $m_{\mathcal{N}} := \min_{i=1,\ldots,n} m_i \geq 2$. Hence, by decreasing c_2' and increasing r, if necessary, we obtain

$$|f(x) - f(x^*)|^{\frac{1}{m_{\mathcal{N}}}} \geq c_2' \|x - x^*\| \quad \text{for } \|x - x^*\| \geq r. \tag{5.7}$$

Since the polynomial function $x \mapsto f(x) - f(x^*)$ is continuous and positive on the compact set $\{x \in \mathbb{R}^n \,|\, \delta \leq \|x\| \leq r\}$, it is not hard to see that

$$|f(x) - f(x^*)|^{\frac{1}{m_{\mathcal{N}}}}$$
$$+ |f(x) - f(x^*)|^{\frac{1}{2}} \geq c_3' \|x - x^*\| \quad \text{for } \delta \leq \|x - x^*\| \leq r$$

for some constant $c_3' > 0$. This inequality, together with inequalities (5.6) and (5.7), implies the second required inequality:

$$|f(x) - f(x^*)|^{\frac{1}{m_{\mathcal{N}}}} + |f(x) - f(x^*)|^{\frac{1}{2}} \geq c_2 \|x - x^*\| \quad \text{for } x \in \mathbb{R}^n,$$

where $c_2 := \min\{c_1, c_2', c_3'\}$. $\qquad \square$

Bibliographic notes

The material in this chapter is taken from a paper by Đoạt, Hà and Phạm in [32].

The first fundamental concept of well-posedness in optimization is inspired by the classical idea of Hadamard [55]. It requires existence and uniqueness of the optimal solution together with continuous dependence on the problem's data. In [145], Tykhonov introduced another concept of well-posedness imposing convergence of every minimizing sequence to the unique minimum point. Its relevance to the approximate (numerical) solution of optimization problems is clear.

Well-posedness for the class of quadratic optimization problems is established by Ioffe, Lucchetti and Revalski in [59]. We refer the reader to the monograph [33] and the survey [60] for basic aspects of the mathematical theory of well-posedness in optimization.

In [104], Morse showed that any smooth function could be approximated by one which has only non-degenerate critical points, and that such a function could be expressed (local near a critical point) as a linear combination of squares of coordinates. We refer the reader to [44, 65, 102] for advances concerning the theory and applications of Morse functions.

The notion of non-degeneracy is introduced by Kouchnirenko in [71]. Theorem 5.2 is a real version of Theorem II(iii) in the cited paper. The proof of the openness presented here is an adaptation of that in [116, Appendix]. The two-sided inequalities in Theorem 5.3 first appeared in the papers by Gindikin [43] and Mikhalov [100].

Chapter 6

Generic Properties in Polynomial Optimization

Abstract

In this chapter, we study genericity for the following parameterized class of nonlinear programs:

$$\text{infimum } f_u(x) := f(x) - \langle u, x \rangle \quad \text{subject to } x \in S,$$

where $f \colon \mathbb{R}^n \to \mathbb{R}$ is a polynomial function and $S \subset \mathbb{R}^n$ is a basic closed semi-algebraic set, which is not necessarily compact. Assume that the constraint set S is regular. It is shown that there exists an open and dense semi-algebraic set $\mathscr{U} \subset \mathbb{R}^n$ such that for any $\bar{u} \in \mathscr{U}$, the restriction of $f_{\bar{u}}$ on S is good at infinity and if $f_{\bar{u}}$ is bounded from below on S, then for all parameters $u \in \mathbb{R}^n$, sufficiently close to $\bar{u}$, the optimization problem $\min_{x \in S} f_u(x)$ has the following properties: the objective function f_u is coercive and has the same growth at infinity on the constraint set S, there is a unique optimal solution, lying on a unique active manifold, for which the strong second-order sufficient conditions, the quadratic growth condition and the global sharp minima hold, and all minimizing sequences converge. Furthermore, the set of active constraints is constant, and the optimal solution and the optimal value function depend analytically under local perturbations of the objective function.

6.1 Regularity

It is well known that the KKT optimality conditions are not actually necessary for optimality. They are only necessary under an additional assumption — a constraint qualification — that in practical is often difficult or impossible to verify. Many constraint qualifications have been studied, but the strongest that ever seems to be needed for the purpose of drawing some conditions about the KKT optimality conditions is regularity of constraint sets. The next result states that this constraint qualification holds generically.

Theorem 6.1. *Let $g_1, \ldots, g_l, h_1, \ldots, h_m$ be polynomials on $\mathbb{R}^n$. Then there exists an open and dense semi-algebraic set $\mathscr{U} \subset \mathbb{R}^l \times \mathbb{R}^m$ such that for any $(a, b) \in \mathscr{U}$, the corresponding set*

$$S(a, b) := \{x \in \mathbb{R}^n \mid g_1(x) = a_1, \ldots, g_l(x) = a_l,$$
$$h_1(x) \geq b_1, \ldots, h_m(x) \geq b_m\}$$

is regular, i.e. for each $x \in S(a, b)$, the gradient vectors $\nabla g_i(x)$, $i = 1, \ldots, l$, and $\nabla h_j(x)$, $j \in \{j \mid h_j(x) = b_j\}$ are linearly independent.

Proof. For each subset $J \subset \{1, \ldots, m\}$, let Φ_J denote the polynomial map

$$\Phi_J \colon \mathbb{R}^n \to \mathbb{R}^l \times \mathbb{R}^{\#J}, \quad x \mapsto (g_i(x), h_j(x))_{i=1,\ldots,l,\ j \in J}.$$

By Sard's theorem (Theorem 1.9), the set Σ_J of critical values of Φ_J is a semi-algebraic set of dimension smaller than $l + \#J$. So, if $(a, b) \mapsto (a, b_J)$ denotes the projection onto $\mathbb{R}^l \times \mathbb{R}^{\#J}$,

$$\mathscr{U}_J := \{(a, b) \in \mathbb{R}^l \times \mathbb{R}^m \mid (a, b_J) \in \mathbb{R}^l \times \mathbb{R}^{\#J} \backslash \overline{\Sigma_J}\}$$

is an open and dense semi-algebraic subset of $\mathbb{R}^l \times \mathbb{R}^m$. It is clear that the set $\mathscr{U} := \bigcap_{J \subset \{1, \ldots, m\}} \mathscr{U}_J$ has the desired properties. $\square$

6.2 Optimality conditions

Let $f, g_1, \ldots, g_l, h_1, \ldots, h_m$ be polynomials on $\mathbb{R}^n$ with degree at most d. We will assume that the closed semi-algebraic set

$$S := \{x \in \mathbb{R}^n \mid g_1(x) = 0, \ldots, g_l(x) = 0,$$
$$h_1(x) \geq 0, \ldots, h_m(x) \geq 0\}$$

is non-empty. For each parameter $u \in \mathbb{R}^n$ let us consider the polynomial $f_u \colon \mathbb{R}^n \to \mathbb{R}, x \mapsto f(x) - \langle u, x \rangle$, and the set

$$\mathrm{KKT}(u) := \Bigg\{ x \in S \mid \text{there exist } \lambda_i, \nu_j \in \mathbb{R} \text{ such that}$$

$$\nabla f_u(x) - \sum_{i=1}^{l} \lambda_i \nabla g_i(x) - \sum_{j=1}^{m} \nu_j \nabla h_j(x) = 0,$$

$$\nu_j h_j(x) = 0, \ \nu_j \geq 0, \ \text{ for } j = 1, \ldots, m \Bigg\}.$$

By definition, $\mathrm{KKT}(u)$ is a subset of the set $\Sigma(f_u, S)$ of critical points of f_u on S.

Lemma 6.1. *There exists an integer $N \geq 1$ such that for every $u \in \mathbb{R}^n$, the set $\mathrm{KKT}(u)$ has at most N connected components. Furthermore, if S is regular, then $\mathrm{KKT}(u)$ is a closed set.*

Proof. For simplicity, we write $\lambda := (\lambda_1, \ldots, \lambda_l) \in \mathbb{R}^l$ and $\nu := (\nu_1, \ldots, \nu_m) \in \mathbb{R}^m$. For each $u \in \mathbb{R}^n$, let

$$W(u) := \Bigg\{ (x, \lambda, \nu) \in \mathbb{R}^n \times \mathbb{R}^l \times \mathbb{R}^m \mid g_i(x) = 0, \ i = 1, \ldots, l,$$

$$h_j(x) \geq 0, j = 1, \ldots, m,$$

$$\nabla f_u(x) - \sum_{i=1}^{l} \lambda_i \nabla g_i(x) - \sum_{j=1}^{m} \nu_j \nabla h_j(x) = 0,$$

$$\nu_j h_j(x) = 0, \ \nu_j \geq 0, \ \text{ for } j = 1, \ldots, m \Bigg\}.$$

Then the set $W(u)$ is semi-algebraic. Note that $\mathrm{KKT}(u) = \pi(W(u))$, where the projection $\pi \colon \mathbb{R}^n \times \mathbb{R}^l \times \mathbb{R}^m \to \mathbb{R}^n$ is defined by $\pi(x, \lambda, \nu) := x$. These observations, together with the Tarski–Seidenberg theorem (Theorem 1.5), yield the first claim. The second claim is an immediately consequence of Theorem 1.4. Finally, the closeness of $\mathrm{KKT}(u)$ follows from the regularity assumption of S. $\qquad \square$

Let $u \in \mathbb{R}^n$ and consider the problem $\min_{x \in S} f_u(x)$. Let $\bar{x}$ be a local minimizer of the problem and $J(\bar{x}) := \{j_1, \ldots, j_k\}$ be the index set of active inequality constraints at $\bar{x}$. If S is regular at $\bar{x}$, i.e. the gradient vectors $\nabla g_i(\bar{x})$, $i = 1, \ldots, l$, and $\nabla h_j(\bar{x})$, $j \in J(\bar{x})$, are linearly independent, then by the KKT optimality conditions (Theorem 2.2) there exist (unique) Lagrange multipliers $\lambda_i, \nu_j \in \mathbb{R}$ satisfying

$$\nabla f(\bar{x}) - \sum_{i=1}^{l} \lambda_i \nabla g_i(a) - \sum_{j=1}^{m} \nu_j \nabla h_j(a) = 0,$$

$$\nu_j h_j(\bar{x}) = 0, \quad \nu_j \geq 0, \quad \text{for } j = 1, \ldots, m.$$

The second equation is called the *complementarity condition*. If it further holds that

$$\nu_1 + h_1(\bar{x}) > 0, \ldots, \nu_m + h_m(\bar{x}) > 0$$

we say the *strict complementarity condition* holds at $\bar{x}$. Let $L(x)$ be the associated Lagrangian function

$$L(x) := f(x) - \sum_{i=1}^{l} \lambda_i g_i(x) - \sum_{j \in J(\bar{x})} \nu_j h_j(x).$$

Then, under the regularity condition, the *second-order necessity condition* holds at $\bar{x}$, i.e.

$$v^T \nabla^2 L(\bar{x}) v \geq 0 \quad \text{for all } v \in T_{\bar{x}} S.$$

Here $\nabla^2 L(\bar{x})$ is the Hessian of L at $\bar{x}$ and $T_{\bar{x}} S$ stands for the *(generalized) tangent space* of S at $\bar{x}$:

$$T_{\bar{x}} S := \{v \in \mathbb{R}^n \mid \langle v, \nabla g_i(\bar{x}) \rangle = 0, \ i = 1, \ldots, l,$$

$$\langle v, \nabla h_j(\bar{x}) \rangle = 0, \ j \in J(\bar{x})\}.$$

If it holds that

$$v^T \nabla^2 L(\bar{x}) v > 0 \quad \text{for all } v \in T_{\bar{x}} S, v \neq 0,$$

we say the *second-order sufficiency condition* holds at $\bar{x}$.

The next result shows that for almost parameter $u \in \mathbb{R}^n$, every local (or global) minimizer satisfies the strong form of the optimality conditions.

Theorem 6.2. *Assume that S is regular. Then there exist an integer $N \geq 1$ and an open and dense semi-algebraic set $\mathscr{U}$ in $\mathbb{R}^n$ such that for any $u \in \mathscr{U}$ the following statements hold:*

(i) *The set $\mathrm{KKT}(u)$ is finite and has at most N points.*

(ii) *For the problem $\min_{x \in S} f_u(x)$, the strict complementarity and second-order sufficiency conditions hold at every local (or global) minimizer.*

Proof. (i) For each subset $J := \{j_1, \ldots, j_k\}$ of $\{1, \ldots, m\}$, we let $\tilde{\nu}_J := (\tilde{\nu}_j)_{j \in J} \in \mathbb{R}^{\#J}$ and

$$V_J := \{(x, \lambda, \tilde{\nu}) \in \mathbb{R}^n \times \mathbb{R}^l \times \mathbb{R}^{\#J} \mid h_j(x) > 0, \text{ for } j \notin J\}.$$

Clearly, V_J is an open semi-algebraic set in $\mathbb{R}^n \times \mathbb{R}^l \times \mathbb{R}^{\#J}$. Assume that $V_J \neq \emptyset$. We define the semi-algebraic map $\Phi_J \colon \mathbb{R}^n \times V_J \to \mathbb{R}^n \times \mathbb{R}^l \times \mathbb{R}^{\#J}$ by

$$\Phi_J(u, x, \lambda, \tilde{\nu}_J) := \left(\nabla f_u(x) - \sum_{i=1}^{l} \lambda_i \nabla g_i(x) - \sum_{j \in J} \tilde{\nu}_j^2 \nabla h_j(x), \right.$$

$$\left. g_1(x), \ldots, g_l(x), h_{j_1}(x), \ldots, h_{j_k}(x) \right).$$

A direct computation shows that

$$(D_u \Phi_J \mid D_x \Phi_J) = \begin{pmatrix} -I_n & \cdots \\ \hline 0 & [\nabla g_1(x)]^T \\ \vdots & \vdots \\ 0 & [\nabla g_l(x)]^T \\ 0 & [\nabla h_{j_1}(x)]^T \\ \vdots & \vdots \\ 0 & [\nabla h_{j_k}(x)]^T \end{pmatrix},$$

where $D_u \Phi_J$ (respectively, $D_x \Phi_J$) denotes the derivative of Φ_J with respect to u (respectively, x), and I_n denotes the identity matrix of order n. Since the constraint set S is regular, it follows that $0 \in \mathbb{R}^n \times \mathbb{R}^l \times \mathbb{R}^{\#J}$ is a regular value of Φ_J. By Sard's theorem with parameter (Theorem 1.10), there exists an open and dense semi-algebraic set $\mathscr{U}_J$ in $\mathbb{R}^n$ such that for each $u \in \mathscr{U}_J$, $0 \in \mathbb{R}^n \times \mathbb{R}^l \times \mathbb{R}^{\#J}$ is a regular value of the map

$$\Phi_{J,u} \colon V_J \to \mathbb{R}^n \times \mathbb{R}^l \times \mathbb{R}^{\#J}, \quad (x, \lambda, \tilde{\nu}_J) \mapsto \Phi_J(u, x, \lambda, \tilde{\nu}_J).$$

Let

$$W_J(u) := \{(x, \lambda, \tilde{\nu}_J) \in V_J \mid \Phi_{J,u}(x, \lambda, \tilde{\nu}_J) = 0\} \quad \text{for } u \in \mathbb{R}^n.$$

It follows from the Inverse Function Theorem that for each $u \in \mathscr{U}_J$, all points of $W_J(u)$ are isolated. Note that $W_J(u)$ is a semi-algebraic set; so in view of Theorem 1.13, it has finitely many connected components. Therefore, $W_J(u)$ is a finite (possibly empty) set for each $u \in \mathscr{U}_J$. Furthermore, by the Implicit Function Theorem, all (local) solutions $(x, \lambda, \tilde{\nu}_J)$ of the system $\Phi_{J,u}(x, \lambda, \tilde{\nu}_J) = 0$ depend analytically on the parameter $u \in \mathscr{U}_J$.

Let $\mathscr{U} := \bigcap_J \mathscr{U}_J$, where the intersection is taken over all subsets J of $\{1, \ldots, m\}$. Then $\mathscr{U}$ is an open and dense semi-algebraic set in $\mathbb{R}^n$.

On the other hand, by construction, it is not hard to see that $x \in \mathrm{KKT}(u)$ if and only if $x \in \pi_J(W_J(u))$, where $\pi_J(x, \lambda, \tilde{\nu}_J) := x$ and $J := J(x)$. Therefore,

$$\mathrm{KKT}(u) = \bigcup_J \pi_J(W_J(u)).$$

Consequently, for any $u \in \mathscr{U}$, the set $\mathrm{KKT}(u)$ is a finite set, and so, by Lemma 6.1, it has at most N points.

(ii) Take any $u \in \mathscr{U}$ and let $\bar{x} \in S$ be a local (or global) minimizer of the polynomial f_u on S. Since S is regular, there exist (unique) Lagrange multipliers $\lambda_1, \ldots, \lambda_l$ and $\nu_1, \ldots, \nu_m$ such that

$$\nabla f_u(\bar{x}) - \sum_{i=1}^{l} \lambda_i \nabla g_i(\bar{x}) - \sum_{j=1}^{m} \nu_j \nabla h_j(\bar{x}) = 0,$$

$$\nu_j h_j(\bar{x}) = 0, \quad \nu_j \geq 0, \quad \text{for } j = 1, \ldots, m.$$

Let $L(x)$ be the associated Lagrangian function

$$L(x) := f_u(x) - \sum_{i=1}^{l} \lambda_i g_i(x) - \sum_{j \in J(\bar{x})} \nu_j h_j(x).$$

Since the constraint set S is regular, the second-order necessary condition holds at $\bar{x}$, i.e.

$$v^T \nabla^2 L(\bar{x}) v \geq 0 \quad \text{for all } v \in T_{\bar{x}} S, \ v \neq 0.$$

We will show that the above inequality is strict.

By contradiction, suppose that there exists a non-zero vector $v \in T_{\bar{x}} S$ such that $v^T \nabla^2 L(\bar{x}) v = 0$. This implies that v is a minimizer of the optimization problem

$$\min_{z \in \mathbb{R}^n} z^T \nabla^2 L(\bar{x}) z \quad \text{such that } \mathcal{M}(\bar{x}) z = 0,$$

where $\mathcal{M}(\bar{x})$ stands for the Jacobian of the active constraint functions

$$\mathcal{M}(\bar{x}) := [\nabla g_i(\bar{x}), \ i = 1, \ldots, l, \ \nabla h_j(\bar{x}), \ j \in J(\bar{x})]^T.$$

By the first-order optimality condition for the above problem, there exists a vector $w \in \mathbb{R}^n$ such that $\nabla^2 L(\bar{x}) v - \mathcal{M}(\bar{x})^T w = 0$, which then implies

$$\begin{pmatrix} \nabla^2 L(\bar{x}) & \mathcal{M}(\bar{x})^T \\ \mathcal{M}(\bar{x}) & 0 \end{pmatrix} \begin{pmatrix} v \\ -w \end{pmatrix} = 0.$$

Since $v \neq 0$, it follows that

$$\det \begin{pmatrix} \nabla^2 L(\bar{x}) & \mathcal{M}(\bar{x})^T \\ \mathcal{M}(\bar{x}) & 0 \end{pmatrix} = 0. \tag{6.1}$$

Now let $J := J(\bar{x})$, $\lambda := (\lambda_1, \ldots, \lambda_l) \in \mathbb{R}^l$, and $\tilde{\nu}_J := (\sqrt{\nu_j})_{j \in J} \in \mathbb{R}^{\#J}$. Since $u \in \mathscr{U}$, we have $0 = \Phi_{J,u}(\bar{x}, \lambda, \tilde{\nu}_J)$ is a regular value of the map $\Phi_{J,\bar{u}}(\cdot, \cdot, \cdot)$. This contradicts (6.1).

Finally, we show that $\nu_j > 0$ for all $j \in J(\bar{x})$. To see this, let us write $J(\bar{x}) := \{j_1, \ldots, j_k\}$ with $1 \leq j_1 < j_2 < \cdots < j_k \leq m$. It is easy to see that if for some $j_\ell \in J(\bar{x})$ we have $\lambda_{j_\ell} = 0$, then the $(n + \ell)$th column of the Jacobian of the map $\Phi_{J,u}(\cdot, \cdot, \cdot)$ at $(\bar{x}, \lambda, \tilde{\nu}_J)$ will vanish, in contradiction to non-singularity. $\qquad \square$

Remark 6.1. It is not hard to see that for each $\bar{u} \in \mathscr{U}$, there exist $\epsilon > 0$ and $N := \#\text{KKT}(\bar{u})$ analytic maps

$$x^i : \{u \in \mathbb{R}^n \mid \|u - \bar{u}\| < \epsilon\} \to \mathbb{R}^n, \quad u \mapsto x^i(u),$$

for $i = 1, \ldots, N$, such that $x^i(u) \neq x^j(u)$, with $i \neq j$, and $\text{KKT}(u) = \{x^1(u), \ldots, x^N(u)\}$.

On the other hand, connected semi-algebraic sets are semi-algebraically arcwise connected (see Theorem 1.13). Hence, the number of points of $\mathrm{KKT}(u)$ is constant on each connected component of $\mathscr{U}$.

6.3   Uniqueness of optimal solutions

Let $f, g_1, \ldots, g_l, h_1, \ldots, h_m$ be polynomials on $\mathbb{R}^n$. Assume that the closed semi-algebraic set
$$S := \{x \in \mathbb{R}^n \mid g_1(x) = 0, \ldots, g_l(x) = 0, \ h_1(x) \geq 0, \ldots, h_m(x) \geq 0\}$$
is non-empty. For each parameter $u \in \mathbb{R}^n$ we define the polynomial f_u by
$$f_u(x) := f(x) - \langle u, x \rangle \quad \text{for } x \in \mathbb{R}^n.$$
The next result states that generically the problem $\inf_{x \in S} f_u$ has at most a global minimizer.

Theorem 6.3. *Assume that the constraint set S is regular. Then there exists an open and dense semi-algebraic set $\mathscr{U} \subset \mathbb{R}^n$ such that for each parameter $u \in \mathscr{U}$, the problem $\inf_{x \in S} f_u$ has at most an optimal solution.*

Proof. This is a direct consequence of Theorem 6.2 and of Lemma 6.2 below.   □

Lemma 6.2. *Under the assumptions of Theorem 6.3, there exists an open and dense semi-algebraic set $\mathscr{U} \subset \mathbb{R}^n$ such that for each parameter $u \in \mathscr{U}$, the corresponding polynomial f_u has distinct values on the set*
$$\Big\{ x \in \mathbb{R}^n \mid \text{there exist } (\lambda_1, \ldots, \lambda_l) \in \mathbb{R}^l \text{ and } (\nu_1, \ldots, \nu_m) \in \mathbb{R}^m$$
$$\text{such that } \nabla f_u(x) - \sum_{i=1}^{l} \lambda_i \nabla g_i(x) - \sum_{j=1}^{m} \nu_j \nabla h_j(x) = 0,$$
$$g_i(x) = 0, \quad \text{for } i = 1, \ldots, l,$$
$$\nu_j h_j(x) = 0, \quad \text{and } \nu_j + [h_j(x)]^2 > 0, \text{ for } j = 1, \ldots, m \Big\}.$$

Proof. Let $\mathscr{L}\colon \mathbb{R}^n \times \mathbb{R}^n \times \mathbb{R}^l \times \mathbb{R}^m \to \mathbb{R}$ be a polynomial function defined by

$$\mathscr{L}(u, x, \lambda, \tilde{\nu}) := f_u(x) - \sum_{i=1}^{l} \lambda_i g_i(x) - \sum_{j=1}^{m} \tilde{\nu}_j^2 h_j(x).$$

We define the semi-algebraic map $\Phi\colon \mathbb{R}^n \times V \to \mathbb{R}^n \times \mathbb{R}^l \times \mathbb{R}^m$ by

$$\Phi(u, x, \lambda, \tilde{\nu})$$
$$:= (\nabla_x \mathscr{L}(u, x, \lambda, \tilde{\nu}), \nabla_\lambda \mathscr{L}(u, x, \lambda, \tilde{\nu}), \nabla_{\tilde{\nu}} \mathscr{L}(u, x, \lambda, \tilde{\nu})),$$

where

$$V := \{(x, \lambda, \tilde{\nu}) \in \mathbb{R}^n \times \mathbb{R}^l \times \mathbb{R}^m \mid \tilde{\nu}_j^2$$
$$+ [h_j(x)]^2 > 0 \text{ for } j = 1, \ldots, m\}.$$

A direct computation shows that

$$(D_u \Phi \mid D_x \Phi \mid D_{\tilde{\nu}} \Phi)$$

$$= (-1) \times \left(\begin{array}{c|ccc|ccc}
I_n & \cdots & & \cdot & \cdot & & \cdot \\
\hline
0 & [\nabla g_1(x)]^T & & 0 & \cdots & & 0 \\
\vdots & \vdots & & \vdots & \ddots & & \vdots \\
0 & [\nabla g_l(x)]^T & & 0 & \cdots & & 0 \\
\hline
0 & 2\tilde{\nu}_1 [\nabla h_1(x)]^T & & 2h_1(x) & \cdots & & 0 \\
\vdots & \vdots & & \vdots & \ddots & & \vdots \\
0 & 2\tilde{\nu}_m [\nabla h_m(x)]^T & & 0 & & \cdots & 2h_m(x)
\end{array} \right).$$

Since the constraint set S is regular, $0 \in \mathbb{R}^n \times \mathbb{R}^l \times \mathbb{R}^m$ is a regular value of Φ. By Sard's theorem with parameter (Theorem 1.10), there exists an open and dense semi-algebraic set $\mathscr{U}'$ in $\mathbb{R}^n$ such that for each $u \in \mathscr{U}'$, $0 \in \mathbb{R}^n \times \mathbb{R}^l \times \mathbb{R}^m$ is a regular value of the map

$$\Phi_u\colon V \to \mathbb{R}^n \times \mathbb{R}^l \times \mathbb{R}^m, \quad (x, \lambda, \tilde{\nu}) \mapsto \Phi(u, x, \lambda, \tilde{\nu}).$$

Next we define the semi-algebraic map

$$\Psi\colon \mathscr{U}' \times ((V \times V) \backslash \Delta) \to \mathbb{R} \times \mathbb{R}^n \times \mathbb{R}^l \times \mathbb{R}^m \times \mathbb{R}^n \times \mathbb{R}^l \times \mathbb{R}^m$$

by

$$\Psi(u, x, \lambda, \tilde{\nu}, x', \lambda', \tilde{\nu}')$$
$$:= \big(\mathscr{L}(u, x, \lambda, \tilde{\nu}) - \mathscr{L}(u, x', \lambda', \tilde{\nu}'), \Phi(u, x, \lambda, \tilde{\nu}), \Phi(u, x', \lambda', \tilde{\nu}')\big),$$

where we put

$$\Delta := \{(x, \lambda, \tilde{\nu}, x', \lambda', \tilde{\nu}') \in \mathbb{R}^n \times \mathbb{R}^l \times \mathbb{R}^m \times \mathbb{R}^n \times \mathbb{R}^l$$
$$\times \mathbb{R}^m \mid x = x'\}.$$

A direct computation shows that

$$D\Psi = (D_u \Psi \mid D_{(x,\lambda,\tilde{\nu})} \Psi \mid D_{(x',\lambda',\tilde{\nu}')} \Psi)$$
$$= \begin{pmatrix} x - x' & \Phi(u, x, \lambda, \tilde{\nu}) & -\Phi(u, x', \lambda', \tilde{\nu}') \\ \bullet & D_{(x,\lambda,\tilde{\nu})} \Phi(u, x, \lambda, \tilde{\nu}) & 0 \\ \bullet & 0 & D_{(x,\lambda,\tilde{\nu})} \Phi(u, x', \lambda', \tilde{\nu}') \end{pmatrix}.$$

Let $(x, \lambda, \tilde{\nu}, x', \lambda', \tilde{\nu}') \in (V \times V) \backslash \Delta$ and $u \in \mathscr{U}'$ be such that

$$\Phi(u, x, \lambda, \tilde{\nu}) = \Phi(u, x', \lambda', \tilde{\nu}') = 0.$$

Note that $0 \in \mathbb{R}^n \times \mathbb{R}^l \times \mathbb{R}^m$ is a regular value of $\Phi_u(\cdot, \cdot, \cdot)$. Therefore, $x - x' \neq 0$, and

$$\operatorname{rank} D_{(x,\lambda,\tilde{\nu})} \Phi(u, x, \lambda, \tilde{\nu})$$
$$= \operatorname{rank} D_{(x,\lambda,\tilde{\nu})} \Phi(u, x', \lambda', \tilde{\nu}') = n + l + m.$$

Consequently, $0 \in \mathbb{R} \times \mathbb{R}^{2n+2l+2m}$ is a regular value of Ψ. By Sard's theorem with parameter (Theorem 1.10), there exists an open and dense semi-algebraic set $\mathscr{U} \subset \mathscr{U}'$ in $\mathbb{R}^n$ such that for each $u \in \mathscr{U}$, 0 is a regular value of the map

$$\Psi_u \colon (V \times V) \backslash \Delta \to \mathbb{R} \times \mathbb{R}^n \times \mathbb{R}^l \times \mathbb{R}^m \times \mathbb{R}^n \times \mathbb{R}^l \times \mathbb{R}^m$$
$$(x, \lambda, \tilde{\nu}, x', \lambda', \tilde{\nu}') \mapsto \Psi(u, x, \lambda, \tilde{\nu}, x', \lambda', \tilde{\nu}').$$

On the other hand, it is clear that

$$\dim((V \times V) \backslash \Delta) = 2n + 2l + 2m$$
$$< 1 + 2n + 2l + 2m$$
$$= \dim(\mathbb{R} \times \mathbb{R}^n \times \mathbb{R}^l \times \mathbb{R}^m \times \mathbb{R}^n \times \mathbb{R}^l \times \mathbb{R}^m).$$

Therefore, $\Psi_u^{-1}(0) = \emptyset$; in other words, the following system has no solution in $V \times V$:

$$x \neq x', \quad \mathscr{L}(u, x, \lambda, \tilde{\nu}) = \mathscr{L}(u, x', \lambda', \tilde{\nu}'),$$

$$\Phi(u, x, \lambda, \tilde{\nu}) = \Phi(u, x', \lambda', \tilde{\nu}') = 0;$$

or equivalently, the following system has no solution in $V \times V$:

$$x \neq x', \quad f_u(x) = f_u(x'), \quad \Phi(u, x, \lambda, \tilde{\nu}) = \Phi(u, x', \lambda', \tilde{\nu}') = 0.$$

Consequently, the polynomial f_u has distinct values on the set

$$\{x \in \mathbb{R}^n \mid \text{there exist } \lambda \in \mathbb{R}^l \text{ and } \tilde{\nu} \in \mathbb{R}^m, \text{ such that}$$

$$(x, \lambda, \tilde{\nu}) \in V, \ \Phi(u, x, \lambda, \tilde{\nu}) = 0\},$$

which completes the proof. $\qquad\qquad\qquad\qquad\qquad\qquad\qquad\qquad\square$

Remark 6.2. Assume that the constraint set S is regular. We can verify that there exists an open and dense semi-algebraic set $\mathscr{U}$ in $\mathbb{R}^n$ such that for any $u \in \mathscr{U}$, the restriction $f_u|_S$ of f_u on S is a *Morse function* in the sense that the following conditions are satisfied:

(i) The set $\Sigma(f_u, S)$ of critical points of f_u on S is finite.
(ii) The f_u has distinct values on the set $\Sigma(f_u, S)$.
(iii) For each $x \in \Sigma(f_u, S)$ there exist unique real numbers λ_i, $i = 1, \ldots, l$, and ν_j, $j = 1, \ldots, m$, with $\nu_j h_j(x) = 0$, such that $\nabla L(x) = 0$ and the Hessian $\nabla^2 L(x)$ of L at x is non-singular, where

$$L(z) := f_u(z) - \sum_{i=1}^{l} \lambda_i g_i(z) - \sum_{j=1}^{m} \nu_j h_j(z) \quad \text{for } z \in \mathbb{R}^n.$$

The proof of these claims is similar to what we did in Theorem 6.2(i) and Lemma 6.2 and we leave it to the reader to carry out the details.

6.4 Goodness at infinity

Let $f, g_1, \ldots, g_l, h_1, \ldots, h_m$ be polynomials on $\mathbb{R}^n$. Assume that the closed semi-algebraic set

$$S := \{x \in \mathbb{R}^n \mid g_1(x) = 0, \ldots, g_l(x) = 0,$$

$$h_1(x) \geq 0, \ldots, h_m(x) \geq 0\}$$

is non-empty.

The following result states that for a generic parameter u, the restriction of the polynomial $f_u(x) := f(x) - \langle u, x \rangle$ on S has a "good behavior at infinity".

Theorem 6.4. *Assume that the constraint set S is regular. Then there exists an open and dense semi-algebraic set $\mathscr{U} \subset \mathbb{R}^n$ such that for each parameter $u \in \mathscr{U}$, the restriction of f_u on S is good at infinity in the sense that there exist constants $c > 0$ and $R > 0$ such that for each $x \in S$ with $\|x\| \geq R$, we have*

$$\left\| \nabla f_u(x) - \sum_{i=1}^{l} \lambda_i \nabla g_i(x) - \sum_{j=1}^{m} \nu_j \nabla h_j(x) \right\| \geq c$$

for all $\lambda \in \mathbb{R}^l$ and all $\nu \in \mathbb{R}^m$ with $\nu_j h_j(x) = 0$ for $j = 1, \ldots, m$.

Proof. For each subset J of $\{1, \ldots, m\}$, let

$$S_J := \{x \in \mathbb{R}^n \mid g_i(x) = 0 \text{ for } i = 1, \ldots, l, \ h_j(x) = 0, \text{ for } j \in J,$$

$$h_j(x) > 0 \text{ for } j \notin J\}.$$

By definition, S_J is a semi-algebraic set in $\mathbb{R}^n$ and $S = \bigcup_J S_J$.

Let J be a subset of $\{1, \ldots, m\}$ such that $S_J \neq \emptyset$. We define the semi-algebraic map $\mathscr{F}_J \colon S_J \times \mathbb{R}^l \times \mathbb{R}^{\#J} \to \mathbb{R}^n$ by

$$\mathscr{F}_J(x, \lambda, \nu_J) := \nabla f(x) - \sum_{i=1}^{l} \lambda_i \nabla g_i(x) - \sum_{j \in J} \nu_j \nabla h_j(x),$$

where $\lambda := (\lambda_1, \ldots, \lambda_n) \in \mathbb{R}^l$ and $\nu_J := (\nu_j)_{j \in J} \in \mathbb{R}^{\#J}$. Let

$$\mathscr{L}_J := \left\{ u \in \mathbb{R}^n \mid \exists \{(x^k, \lambda^k, \nu_J^k)\}_{k \in \mathbb{N}} \subset S_J \times \mathbb{R}^l \times \mathbb{R}^{\#J} \text{ such that} \right.$$

$$\left. \lim_{k \to \infty} \|(x^k, \lambda^k, \nu_J^k)\| = +\infty \text{ and } \lim_{k \to \infty} \mathscr{F}_J(x^k, \lambda^k, \nu_J^k) = u \right\}.$$

We will show that $\mathscr{L}_J$ is a semi-algebraic set of dimension at most $n - 1$.

To see this, let $\overline{G}$ be the closure of the set

$$G := \{(u, x, \lambda, \nu_J) \in \mathbb{R}^n \times S_J \times \mathbb{R}^l \times \mathbb{R}^{\#J} \mid u = \mathscr{F}_J(x, \lambda, \nu_J)\}$$

$$\subset \mathbb{R}^n \times \mathbb{R}^N.$$

in $\mathbb{R}^n \times \mathbb{P}^N$, where $N := n + l + \#J$ and $\mathbb{P}^N$ is the real projective space. Then, the sets G and $\overline{G}$ are semi-algebraic. Hence, the set $\overline{G} \backslash G$ is semi-algebraic. Moreover, if $\pi \colon \mathbb{R}^n \times \mathbb{P}^N \to \mathbb{R}^n$ is the projection on the first factor, then $\mathscr{Z}_J = \pi(\overline{G} \backslash G)$. In view of the Tarski–Seidenberg theorem (Theorem 1.5), the set $\mathscr{Z}_J$ is semi-algebraic. It follows from Proposition 1.4 that

$$\dim \mathscr{Z}_J = \dim \pi(\overline{G} \backslash G)$$
$$\leq \dim(\overline{G} \backslash G) < \dim G = \dim(S_J \times \mathbb{R}^l \times \mathbb{R}^{\#J}).$$

On the other hand, since the constraint set S is regular, it follows easily that

$$\dim S_J = n - l - \#J.$$

Therefore, $\dim \mathscr{Z}_J \leq n - 1$.

Finally, let $\mathscr{U} := \mathbb{R}^n \backslash \bigcup_J \mathscr{Z}_J$, where the union is taken over all subsets J of $\{1, \ldots, m\}$ for which S_J is non-empty. It is easy to see that $\mathscr{U}$ has the desired properties. $\qquad \square$

6.5 Coercivity

Let $f, g_1, \ldots, g_l, h_1, \ldots, h_m$ be polynomials on $\mathbb{R}^n$. Assume that the closed semi-algebraic set

$$S := \{x \in \mathbb{R}^n \mid g_1(x) = 0, \ldots, g_l(x) = 0,$$
$$h_1(x) \geq 0, \ldots, h_m(x) \geq 0\}$$

is non-empty. For each parameter $u \in \mathbb{R}^n$ we define the polynomial f_u by

$$f_u(x) := f(x) - \langle u, x \rangle \quad \text{for } x \in \mathbb{R}^n.$$

The next result shows that for a generic parameter u, the restriction of f_u on S is coercive provided that f_u is bounded from below on S.

Theorem 6.5. *Assume that the constraint set S is regular. Let $\mathscr{U} \subset \mathbb{R}^n$ be the open and dense semi-algebraic set given in Theorem 6.4. For each parameter $u \in \mathscr{U}$, if the corresponding polynomial f_u is* *bounded from below on S, then there exist constants $c > 0$ and $R > 0$ such that*

$$f_u(x) \geq c\|x\| \quad \text{for all } x \in S \text{ and } \|x\| \geq R.$$

In particular, f_u is coercive on S.

Proof. Take any $u \in \mathcal{U}$ and assume that the polynomial $f_u \colon \mathbb{R}^n \to \mathbb{R}, x \mapsto f_u(x) := f(x) - \langle u, x \rangle$, is bounded from below on S. We will show that there exist constants $c > 0$ and $R > 0$ such that

$$f_u(x) \geq c\|x\| \quad \text{for all } x \in S \text{ and } \|x\| \geq R.$$

Suppose, by contradiction, that there exist sequences $\{x^k\}_{k\in\mathbb{N}} \subset S$, with $\lim_{k\to\infty} \|x^k\| = +\infty$, and $\{c^k\}_{k\in\mathbb{N}} \subset \mathbb{R}$, with $c^k > 0$ and $\lim_{k\to\infty} c^k = 0$, such that

$$f_u(x^k) < c^k\|x^k\| \quad \text{for all } k.$$

For each k, let $y^k \in S$ be a minimizer of the following problem

$$\min_{x\in S,\ \|x\|^2=\|x^k\|^2} f_u(x).$$

(The existence of y^k follows direct from the fact that the objective function f_u is continuous on the compact set $\{x \in S, \|x\|^2 = \|x^k\|^2\}$.) Then $\lim_{k\to\infty} \|y^k\| = +\infty$ and we have for all k,

$$-\infty < \inf_{x\in S} f_u(x) \leq f_u(y^k) \leq f_u(x^k) < c^k\|x^k\| = c^k\|y^k\|,$$

which yields that

$$\lim_{k\to+\infty} \frac{f_u(y^k)}{\|y^k\|} = 0.$$

By Lemma 2.1, for all $k \gg 1$, the vectors $\nabla g_i(y^k), i = 1, \ldots, l$, $\nabla h_j(y^k), j \in J(y^k)$, and the vector y^k are linearly independent. We therefore deduce from the KKT optimality conditions (Theorem 2.2) that: there exist sequences $\{\lambda^k\}_{k\in\mathbb{N}} \subset \mathbb{R}^l, \{\nu^k\}_{k\in\mathbb{N}} \subset \mathbb{R}^m$, and $\{\mu^k\}_{k\in\mathbb{N}} \subset \mathbb{R}$ satisfying the following:

$$\nabla f_u(y^k) - \sum_{i=1}^{l} \lambda_i^k \nabla g_i(y^k) - \sum_{j=1}^{m} \nu_j^k \nabla h_j(y^k) - \mu^k y^k = 0,$$

$$\nu_j^k h_j(y^k) = 0 \quad \text{and} \quad \nu_j^k \geq 0 \text{ for } j = 1, \ldots, m.$$

By the Curve Selection Lemma at infinity (Theorem 1.12), there are a smooth semi-algebraic curve $\varphi(t)$ and semi-algebraic functions $\lambda_i(t), \nu_j(t), \mu(t), t \in (0, \epsilon]$, such that

(a) $\lim_{t \to 0^+} \|\varphi(t)\| = +\infty;$
(b) $\lim_{t \to 0^+} \frac{f_u(\varphi(t))}{\|\varphi(t)\|} = 0;$
(c) $\nabla f_u(\varphi(t)) - \sum_{i=1}^{l} \lambda_i(t) \nabla g_i(\varphi(t)) - \sum_{j=1}^{m} \nu_j(t) \nabla h_j(\varphi(t)) - \mu(t)\varphi(t) \equiv 0;$ and
(d) for all $t \in (0, \epsilon]$, we have $\varphi(t) \in S$ and $\nu_j(t) h_j(\varphi(t)) = 0$, for $j = 1, \dots, m.$

Thanks to the Monotonicity Theorem (Theorem 1.8), for $\epsilon > 0$ small enough, the functions ν_j and $h_j \circ \varphi$ are either constant or strictly monotone. Then, by (d), we can see that either $\nu_j(t) \equiv 0$ or $h_j \circ \varphi(t) \equiv 0$; in particular,

$$\nu_j(t) \frac{d}{dt}(h_j \circ \varphi)(t) \equiv 0, \quad j = 1, \dots, m.$$

Hence, it follows from (c) that

$$\frac{d}{dt}(f_u \circ \varphi)(t) = \left\langle \nabla f_u(\varphi(t)), \frac{d\varphi}{dt} \right\rangle$$

$$= \sum_{i=1}^{l} \lambda_i(t) \left\langle \nabla g_i(\varphi(t)), \frac{d\varphi}{dt} \right\rangle$$

$$+ \sum_{j=1}^{m} \nu_j(t) \left\langle \nabla h_j(\varphi(t)), \frac{d\varphi}{dt} \right\rangle + \mu(t) \left\langle \varphi(t), \frac{d\varphi}{dt} \right\rangle$$

$$= \sum_{i=1}^{l} \lambda_i(t) \frac{d}{dt}(g_i \circ \varphi)(t) + \sum_{j=1}^{m} \nu_j(t) \frac{d}{dt}(h_j \circ \varphi)(t)$$

$$+ \frac{\mu(t)}{2} \frac{d\|\varphi(t)\|^2}{dt}$$

$$= \frac{\mu(t)}{2} \frac{d\|\varphi(t)\|^2}{dt}.$$

This, together with condition (c) again, implies that

$$\left| \frac{d}{dt} (f_u \circ \varphi)(t) \right|$$

$$= \left| \frac{\mu(t)}{2} \frac{d\|\varphi(t)\|^2}{dt} \right|$$

$$= \frac{\|\nabla f_u(\varphi(t)) - \sum_{i=1}^{l} \lambda_i(t) \nabla g_i(\varphi(t)) - \sum_{j=1}^{m} \nu_j(t) \nabla h_j(\varphi(t))\|}{2\|\varphi(t)\|}$$

$$\times \left| \frac{d\|\varphi(t)\|^2}{dt} \right|.$$

Since the function $f_u \circ \varphi \colon (0, \epsilon] \mapsto \mathbb{R}, t \mapsto f_u(\varphi(t))$, is semi-algebraic, it follows from the Monotonicity Theorem (Theorem 1.8) that for $\epsilon > 0$ sufficiently small, this functions is either constant or strictly monotone. If the function $f_u \circ \varphi$ is constant, then $\mu(t) \equiv 0$, and hence

$$\nabla f_u(\varphi(t)) - \sum_{i=1}^{l} \lambda_i(t) \nabla g_i(\varphi(t)) - \sum_{j=1}^{m} \nu_j(t) \nabla h_j(\varphi(t)) \equiv 0,$$

or equivalently,

$$\nabla f(\varphi(t)) - \sum_{i=1}^{l} \lambda_i(t) \nabla g_i(\varphi(t)) - \sum_{j=1}^{m} \nu_j(t) \nabla h_j(\varphi(t)) \equiv u.$$

Combining this equality with condition (a) gives a contradiction to the assumption that $u \in \mathscr{U}$.

Hence, we may assume that the function $(0, \epsilon] \mapsto \mathbb{R}, t \mapsto f_u(\varphi(t))$, is not constant. Thanks to the Growth Dichotomy Lemma (Lemma 1.7), we may write

$$\|\varphi(t)\| = c_\alpha t^\alpha + \text{higher order terms in } t,$$

$$f_u(\varphi(t)) = c_\beta t^\beta + \text{higher order terms in } t,$$

here $c_\alpha \neq 0, c_\beta \neq 0, \alpha < 0$, and $\alpha < \beta$ (because of conditions (a)–(b)). By a direct computation,

$$\left\| \nabla f_u(\varphi(t)) - \sum_{i=1}^{l} \lambda_i(t) \nabla g_i(\varphi(t)) - \sum_{j=1}^{m} \nu_j(t) \nabla h_j(\varphi(t)) \right\|$$

$$= ct^{\beta-\alpha} + o(t^{\beta-\alpha}),$$

for some constant $c \neq 0$, which now yields

$$\lim_{t \to 0^+} \left\| \nabla f_u(\varphi(t)) - \sum_{i=1}^{l} \lambda_i(t) \nabla g_i(\varphi(t)) - \sum_{j=1}^{m} \nu_j(t) \nabla h_j(\varphi(t)) \right\| = 0.$$

Consequently, we get

$$\lim_{t \to 0^+} \left[\nabla f(\varphi(t)) - \sum_{i=1}^{l} \lambda_i(t) \nabla g_i(\varphi(t)) - \sum_{j=1}^{m} \nu_j(t) \nabla h_j(\varphi(t)) \right] = u.$$

This equality and condition (a) give us a contradiction to the fact that $u \in \mathscr{U}$. $\qquad \square$

Remark 6.3. Theorem 6.5 can be proved by using the Ekeland variational principle (Theorem 3.1); the details are left to the reader.

6.6 Genericity in polynomial optimization

Let $f, g_1, \ldots, g_l, h_1, \ldots, h_m$ be polynomials on $\mathbb{R}^n$. Assume that the closed semi-algebraic set

$$S := \{x \in \mathbb{R}^n \mid g_1(x) = 0, \ldots, g_l(x) = 0,$$
$$h_1(x) \geq 0, \ldots, h_m(x) \geq 0\}$$

is non-empty. For each parameter $u \in \mathbb{R}^n$ we define the polynomial f_u by

$$f_u(x) := f(x) - \langle u, x \rangle \quad \text{for } x \in \mathbb{R}^n.$$

With these definitions in hand, generic properties for the class of polynomial optimization problems are summarized in the following statement.

Theorem 6.6. *Assume that the constraint set S is regular. Then there exists an open and dense semi-algebraic set $\mathscr{U} \subset \mathbb{R}^n$ such that for any $\bar{u} \in \mathscr{U}$, if $f_{\bar{u}}$ is bounded from below on S, then there exist positive constants $\epsilon, R,$ and $c_i, i = 1, 2, 3,$ with $\epsilon < c_1,$ such that for any $u \in \mathbb{R}^n,$ with $\|u - \bar{u}\| < \epsilon,$ we have $u \in \mathscr{U}$ and the following* *statements are satisfied:*

(i) [Uniform coercivity and growth at infinity] *For all* $x \in S$, *with* $\|x\| \geq R$, *the following inequalities hold:*

$$f_u(x) := f(x) - \langle u, x \rangle \geq c_1 \|x\|,$$

$$\left(1 - \frac{\epsilon}{c_1}\right)^{-1} f_u(x) \geq f_{\bar{u}}(x) \geq \left(1 - \frac{\epsilon}{c_1}\right) f_u(x).$$

In particular, f_u *is coercive on* S.

(ii) [Existence, uniqueness, and optimality conditions] *The problem* $\min_{x \in S} f_u(x)$ *has a unique minimizer* $x(u) \in S$ *for which the strict complementarity and second-order sufficiency conditions hold.*

(iii) [Analyticity of the optimal solution] *The corresponding*

$$\{u \in \mathbb{R}^n \mid \|u - \bar{u}\| < \epsilon\} \to \mathbb{R}^n \times \mathbb{R}^l \times \mathbb{R}^m,$$

$$u \mapsto (x(u), \lambda(u), \nu(u)),$$

is an analytic map, where $\lambda(u) \in \mathbb{R}^l$ *and* $\nu(u) \in \mathbb{R}^m$ *are the (unique) Lagrange multipliers with respect to* $x(u)$.

(iv) [Analyticity of the optimal value function] *The function*

$$\phi \colon \{u \in \mathbb{R}^n \mid \|u - \bar{u}\| < \epsilon\} \to \mathbb{R}, \quad u \mapsto \min_{x \in S} f_u(x),$$

is analytic and $\nabla \phi(u) = -x(u)$.

(v) [Local constancy of the set of active constraint indices] *The set of active constraint indices is locally constant:*

$$\{j \mid h_j(x(u)) = 0\} = \{j \mid h_j(x(\bar{u})) = 0\}.$$

(vi) [Uniform quadratic growth condition] *We have for any* $x \in S$, *with* $\|x - x(u)\| \leq R$,

$$f_u(x) - f_u(x(u)) \geq c_2 \|x - x(u)\|^2.$$

(vii) [Uniform and global sharp minima] *For any* $x \in S$, *the following inequality holds:*

$$[f_u(x) - f_u(x(u))] + [f_u(x) - f_u(x(u))]^{\frac{1}{2}} \geq c_3 \|x - x(u)\|.$$

(viii) [Convergence of all minimizing sequences] *For any sequence* $\{u^k\} \subset \mathbb{R}^n$ *converging to* $\bar{u}$, $\inf_{x \in S} f_{u^k}(x)$ *is finite for large* k *and any sequence* $\{x^k\} \subset S$ *such that* $f_{u^k}(x^k) - \inf_{x \in S} f_{u^k}(x) \to 0$ *converges to* $x(\bar{u})$.

Proof. Let $\mathscr{U} \subset \mathbb{R}^n$ be an open and dense semi-algebraic set for which the statements of Theorems 6.2–6.5 hold.

Take any $\bar{u} \in \mathscr{U}$. Thanks to Theorem 6.4, the restriction of $f_{\bar{u}}$ on S is good at infinity.

Next we assume that the function $x \mapsto f_{\bar{u}}(x) = f(x) - \langle \bar{u}, x \rangle$ is bounded from below on the constraint set S.

(i) By Theorem 6.5, there exist positive constants c' and R such that

$$f_{\bar{u}}(x) \geq c' \|x\| \quad \text{for all } x \in S, \quad \|x\| \geq R.$$

Since the set $\mathscr{U}$ is open, there exists $\epsilon \in (0, \frac{c'}{2})$ such that $\{u \in \mathbb{R}^n \mid \|u - \bar{u}\| < \epsilon\} \subset \mathscr{U}$. Then we have for all $x \in S$ and $\|x\| \geq R$,

$$f_u(x) = f_{\bar{u}}(x) - \langle u - \bar{u}, x \rangle \geq f_{\bar{u}}(x) - \|u - \bar{u}\| \|x\|$$

$$\geq (c' - \|u - \bar{u}\|) \|x\| > (c' - \epsilon) \|x\| \geq \frac{c'}{2} \|x\|.$$

Clearly, the positive constant $c_1 := \frac{c'}{2}$ satisfies the required inequality. Consequently, the function f_u is coercive on S.

Let $u \in \mathbb{R}^n$ be such that $\|u - \bar{u}\| < \epsilon$. It is clear that

$$|\langle u - \bar{u}, x \rangle| \leq \|u - \bar{u}\| \|x\| \leq \epsilon \|x\|.$$

It follows for all $x \in S$, with $\|x\| \geq R$, that

$$f_u(x) = f(x) - \langle u, x \rangle = f_{\bar{u}}(x) - \langle u - \bar{u}, x \rangle$$

$$\geq f_{\bar{u}}(x) - |\langle u - \bar{u}, x \rangle| \geq f_{\bar{u}}(x) - \epsilon \|x\|$$

$$\geq \left(1 - \frac{\epsilon}{c_1}\right) f_{\bar{u}}(x).$$

Similarly, we also have

$$f_{\bar{u}}(x) \geq \left(1 - \frac{\epsilon}{c_1}\right) f_u(x) \quad \text{for all } x \in S, \quad \|x\| \geq R.$$

The proof of (i) completes.

(ii) Take any $u \in \mathbb{R}^n$, with $\|u - \bar{u}\| < \epsilon$. Then $u \in \mathscr{U}$. By Theorem 6.2, the strict complementarity and second-order sufficiency conditions hold at every local (or global) minimizer of the polynomial f_u on S. Since the polynomial f_u is coercive on S, it has a global minimizer on S. Then it follows from Theorem 6.3 that the optimization problem $\min_{x \in S} f_u(x)$ has a unique minimizer $x(u) \in S$ for which the strict complementarity and second-order sufficiency conditions satisfy.

(iii) We first show that $\lim_{u \to \bar{u}} x(u) = x(\bar{u})$. Indeed, we see that if $\|x(u)\| \geq R$, $\|u - \bar{u}\| < \epsilon$, then

$$c_1 \|x(u)\| \leq f_u(x(u)) \leq f_u(x(\bar{u}))$$
$$= f(x(\bar{u})) - \langle u, x(\bar{u}) \rangle$$
$$= f(x(\bar{u})) - \langle \bar{u}, x(\bar{u}) \rangle - \langle u - \bar{u}, x(\bar{u}) \rangle$$
$$= f_{\bar{u}}(x(\bar{u})) - \langle u - \bar{u}, x(\bar{u}) \rangle$$
$$\leq f_{\bar{u}}(x(\bar{u})) + \|u - \bar{u}\| \|x(\bar{u})\|$$
$$\leq f_{\bar{u}}(x(\bar{u})) + \epsilon \|x(\bar{u})\|.$$

This implies that the set $\{x(u) \mid \|u - \bar{u}\| < \epsilon\}$ is bounded. Let $\bar{y} \in \{\lim_{u \to \bar{u}} x(u)\}$. Then $\bar{y} \in S$. Note that $f_u(x(u)) \leq f_u(x(\bar{u}))$. Consequently,

$$f_{\bar{u}}(\bar{y}) \leq \lim_{u \to \bar{u}} f_u(x(\bar{u})) = \lim_{u \to \bar{u}} [f(x(\bar{u})) - \langle u, x(\bar{u}) \rangle]$$
$$= f(x(\bar{u})) - \langle \bar{u}, x(\bar{u}) \rangle = f_{\bar{u}}(x(\bar{u})),$$

which yields $\bar{y} = x(\bar{u})$ because $x(\bar{u})$ is the unique minimizer of $f_{\bar{u}}$ on S. Therefore, there exists the limit $\lim_{u \to \bar{u}} x(u) = x(\bar{u})$.

Since the constraint set S is regular, for each $u \in \mathbb{R}^n$, with $\|u - \bar{u}\| < \epsilon$, there exist the unique Lagrange multipliers $\lambda(u) \in \mathbb{R}^l$ and $\nu(u) \in \mathbb{R}^m$ corresponding to the minimizer $x(u)$. It is easy to see that $\lim_{u \to \bar{u}} \lambda(u) = \lambda(\bar{u})$ and $\lim_{u \to \bar{u}} \nu(u) = \nu(\bar{u})$.

Keeping the notations as in the proof of Lemma 6.2. Let

$$\tilde{\nu}(u) := (\sqrt{\nu_1(u)}, \ldots, \sqrt{\nu_m(u)}) \in \mathbb{R}^m.$$

We have $(x(u), \lambda(u), \tilde{\nu}(u)) \in V$ and $\Phi_u(x(u), \lambda(u), \tilde{\nu}(u)) = 0$. Since $0 \in \mathbb{R}^n \times \mathbb{R}^l \times \mathbb{R}^m$ is a regular value of the map $\Phi_{\bar{u}}$, the Jacobian

$D_{(x,\lambda,\tilde{\nu})}\Phi_{\bar{u}}(x(\bar{u}),\lambda(\bar{u}),\tilde{\nu}(\bar{u}))$ is non-singular. By the Implicit Function Theorem, the system $\Phi_u(x,\lambda,\tilde{\nu}) = 0$ has a unique solution, which depends analytically on u in some neighborhood of the point $(x(\bar{u}),\lambda(\bar{u}),\tilde{\nu}(\bar{u}))$. This, together with the fact that $\lim_{u\to\bar{u}} x(u) = x(\bar{u})$, implies that if $\epsilon > 0$ small enough, then the map $u \mapsto (x(u),\lambda(u),\tilde{\nu}(u))$ is analytic.

(iv) By definition, we have for all $u \in \mathbb{R}^n$, with $\|u - \bar{u}\| < \epsilon$,

$$\phi(u) = \min_{x\in S} f_u(x) = f_u(x(u)) = ((f \circ x) - \langle u, x\rangle)(u).$$

The map $u \mapsto x(u)$ is analytic, so is $u \mapsto \phi(u)$.

We have shown that for all $u \in \mathbb{R}^n$, with $\|u - \bar{u}\| < \epsilon$,

$$0 = \nabla f(x(u)) - u - \sum_{i=1}^{l}\lambda_i(u)\nabla g_i(x(u))$$

$$- \sum_{j\in J(x(\bar{u}))} \nu_j(u)\nabla h_j(x(u)),$$

$$g_i(x(u)) = 0, \ i = 1,\ldots,l, \quad \text{and} \quad h_j(x(u)) = 0, \ j \in J(x(\bar{u})).$$

It follows successively that

$$0 = \nabla f(x(u))Dx(u) - uDx(u) - \sum_{i=1}^{l}\lambda_i(u)\nabla g_i(x(u))Dx(u)$$

$$- \sum_{j\in J(x(\bar{u}))} \nu_j(u)\nabla h_j(x(u))Dx(u)$$

$$= \nabla(f \circ x)(u) - uDx(u) - \sum_{i=1}^{l}\lambda_i(u)\nabla(g_i \circ x)(u)$$

$$- \sum_{j\in J(x(\bar{u}))} \nu_j(u)\nabla(h_j \circ x)(u)$$

$$= \nabla(f \circ x)(u) - uDx(u).$$

Hence

$$\nabla\phi(u) = \nabla(f_u \circ x)(u) = \nabla\left((f \circ x) - \langle u, x\rangle\right)(u)$$

$$= \nabla(f \circ x)(u) - uDx(u) - x(u) = -x(u).$$

This equality proves (iv).

(v) We next show that

$$J(x(u)) = J(x(\bar{u})) \quad \text{for all} \quad u \text{ near } \bar{u}.$$

Indeed, if $j \notin J(x(\bar{u}))$ then $h_j(x(\bar{u})) > 0$, and by continuity, we have for all u near $\bar{u}$, $h_j(x(u)) > 0$ and hence $J(x(u)) \subseteq J(x(\bar{u}))$. Furthermore, the equality $J(x(u)) = J(x(\bar{u}))$ holds for all u near $\bar{u}$. Indeed, if it is not the case, then there exist a sequence $\{u^k\}_{k \in \mathbb{N}} \subset \mathscr{U}$, with $\lim_{k \to +\infty} u^k = \bar{u}$, and an index $j \in J(x(\bar{u})) \backslash J(x(u^k))$. Then $h_j(x(u^k)) > 0$. The complementarity condition implies that $\nu_j(u^k) = 0$ for all k. By continuity, we get $\nu_j(\bar{u}) = 0$, which contradicts the facts that $\nu_j(\bar{u}) + h_j(x(\bar{u})) > 0$ and $h_j(x(\bar{u})) = 0$ (because $j \in J(x(\bar{u}))$). Therefore, (iv) holds.

(vi) The function $\phi(\cdot)$ is differentiable of class C^2 on $\{u \in \mathbb{R}^n \mid \|u - \bar{u}\| < \epsilon\}$. By Taylor expansion, we have for any fixed parameter $u \in \mathbb{R}^n$, with $\|u - \bar{u}\| < \epsilon$,

$$\phi(v) = \phi(u) + \langle v - u, \nabla\phi(u)\rangle + \frac{1}{2}\langle v - u, \nabla^2\phi(u)(v - u)\rangle$$
$$+ o(\|v - u\|^2)$$
$$= \phi(u) - \langle v - u, x(u)\rangle + \frac{1}{2}\langle v - u, \nabla^2\phi(u)(v - u)\rangle$$
$$+ o(\|v - u\|^2)$$

for all $v \in \mathscr{U}$ near u, where $\nabla^2\phi(u)$ denotes the Hessian of the function ϕ at u. Since the Hessian $\nabla^2\phi(\cdot)$ is continuous on $\{u \in \mathbb{R}^n \mid \|u - \bar{u}\| < \epsilon\}$, shrinking $\epsilon > 0$ if necessary we may assume that $\nabla^2\phi(v)$ is bounded for all v satisfying $\|v - \bar{u}\| < \epsilon$. Hence there exist positive constants $\delta < \epsilon - \|u - \bar{u}\|$ and ρ such that

$$\phi(v) \geq \phi(u) - \langle v - u, x(u)\rangle - \frac{\rho}{2}\|v - u\|^2 \quad \text{for all } \|v - u\| < \delta.$$

Furthermore, since the set $S_R := \{x \in S \mid \|x - x(u)\| \leq R\}$ is compact, we can clearly assume that

$$\delta^{-1} \times \max_{x,y \in S_R} \|x - y\| < \rho. \tag{6.2}$$

Now consider any point $x \in S_R$. Since $\phi(v) \leq f_v(x)$ for all $v \in \mathbb{R}^n$, we deduce successively

$$0 \leq \inf_{v \in \mathbb{R}^n} \{f_v(x) - \phi(v)\}$$

$$\leq \inf_{\|v-u\|<\delta} \{f_v(x) - \phi(v)\}$$

$$\leq \inf_{\|v-u\|<\delta} \left\{ f_v(x) - \phi(u) + \langle v - u, x(u) \rangle + \frac{\rho}{2}\|v - u\|^2 \right\}$$

$$= \inf_{\|v-u\|<\delta} \left\{ f_v(x) - f_u(x(u)) + \langle v - u, x(u) \rangle + \frac{\rho}{2}\|v - u\|^2 \right\}$$

$$= \inf_{\|v-u\|<\delta} \left\{ f(x) - \langle v, x \rangle - f_u(x(u)) + \langle v - u, x(u) \rangle + \frac{\rho}{2}\|v - u\|^2 \right\}$$

$$= \inf_{\|v-u\|<\delta} \left\{ f_u(x) - f_u(x(u)) - \langle v - u, x - x(u) \rangle + \frac{\rho}{2}\|v - u\|^2 \right\}$$

$$= f_u(x) - f_u(x(u)) + \inf_{\|v-u\|<\delta} \left\{ -\langle v - u, x - x(u) \rangle + \frac{\rho}{2}\|v - u\|^2 \right\}.$$

It follows easily from inequality (6.2) that the above infimum is attained at the point $v = u + \rho^{-1}(x - x(u))$ satisfying $\|v - u\| < \delta$. Replacing this value in the above inequality, we deduce for all $x \in S_R$ that

$$0 \leq f_u(x) - f_u(x(u)) - \frac{1}{2\rho}\|x - x(u)\|^2,$$

which yields the desired conclusion with $c_2 := (2\rho)^{-1}$.

(vii) By (i), we can find positive constants c_1' and $R' \geq R$ such that

$$[f_u(x) - f_u(x(u))] \geq c_1'\|x - x(u)\| \quad \text{for } x \in S, \ \|x - x(u)\| \geq R'.$$

By applying (vi) we obtain the inequality

$$[f_u(x) - f_u(x(u))]^{\frac{1}{2}} \geq c_2'\|x - x(u)\| \quad \text{for } x \in S, \ \|x - x(u)\| \leq R$$

where $c_2' := \sqrt{c_2} > 0$.

On the other hand, it is not hard to show that

$$[f_u(x) - f_u(x(u))]$$
$$\geq c_3'\|x - x(u)\| \quad \text{for } x \in S, \ R \leq \|x - x(u)\| \leq R'$$

for some $c_3' > 0$.

From the above inequalities we obtain

$$[f_u(x) - f_u(x(u))] + [f_u(x) - f_u(x(u))]^{\frac{1}{2}}$$
$$\geq c\|x - x(u)\| \quad \text{for } x \in S,$$

where $c_3 := \min\{c_1', c_2', c_3'\}$. This implies the required inequality.

(viii) Let $\{u^k\} \subset \mathbb{R}^n$ be a sequence converging to $\bar{u}$. Then, for all sufficiently large k, $\|u^k - \bar{u}\| < \epsilon$, and hence the optimal value $\inf_{x \in S} f_{u^k}(x) = f_{u^k}(x(u^k))$ is finite.

Take any sequence $\{x^k\} \subset S$ with $f_{u^k}(x^k) - \inf_{x \in S} f_{u^k}(x) \to 0$. By (vii), then for all k large enough,

$$[f_{u^k}(x^k) - f_{u^k}(x(u^k))] + [f_{u^k}(x^k) - f_{u^k}(x(u^k))]^{\frac{1}{2}}$$
$$\geq c_3\|x^k - x(u^k)\|.$$

This implies clearly that $\lim_{k \to \infty} x^k = \lim_{k \to \infty} x(u^k) = x(\bar{u})$.

The proof of the theorem completes. $\qquad\qquad\qquad\qquad\qquad\square$

Bibliographic notes

Most results in this chapter are taken from papers [84, 85] by Lee and Phạm.

Theorems 6.1 and 6.2(ii) are due to Spingarn and Rockafellar [142]. Theorem 6.4 can be considered as the real counterpart of [17, Proposition 3.2], where the author has proved it for a single complex polynomial function; the idea of our proof is taken from [62].

The idea to study mathematical programming problems from the generic point of view goes back to the investigation of Saigal and Simon [135] for the complementarity problem. The studies of generic strict complementarity and primal and dual non-degeneracy for semidefinite programming by Alizadeh, Haeberly, and Overton [2] and Shapiro [140]. The study of generic properties of general conic convex programs was given by Pataki and Tunçel [124].

Generic optimality conditions for parameterized classes of nonlinear programming problems are treated by Spingarn and Rockafellar [142] (see also [40, 41, 61, 114, 143]).

It was shown in [53] by Hà and Phạm that almost every linear objective function, which is bounded from below on a closed semi-algebraic set, attains its infimum and has the same asymptotic growth at infinity.

Bolte, Daniilidis, and Lewis [16] established some generic properties for problems of minimizing a linear function over a compact convex semialgebraic set; for refinements of this result see [35, 87]. Nie [114] proved that classical optimality conditions hold in a Zariski open set in the space of input polynomials with given degrees.

Chapter 7

Optimization of Polynomials on Compact Semi-algebraic Sets

Abstract

We consider the problem of minimizing a polynomial over a compact basic semi-algebraic set which is NP-hard in general. By using sums of squares certificates for positive polynomials, we will construct a sequence of semidefinite programs whose optimal values converge monotonically, increasing to the optimal value of the original problem. We show that the sequence has finite convergence provided that either strong optimality conditions hold at every global minimizer or the constraint set is finite.

7.1 Semidefinite programming and sums of squares

This section presents some preliminaries on semidefinite programming and sums of squares of polynomials. It will be shown that the problem of deciding when a polynomial is a sum of squares is a semidefinite program.

7.1.1 *Semidefinite programming*

Denote by $\mathrm{Sym}(s)$ the set of symmetric $s \times s$ matrices. The *trace* of a square matrix is defined to be the sum of the diagonal entries. The standard scalar product on $\mathrm{Sym}(s)$ is defined by $\langle A, B \rangle := \mathrm{Tr}(A^T B)$.

A matrix $A \in \mathrm{Sym}(s)$ is said to be *positive semidefinite* if $\langle v, Av \rangle \geq 0$ for all $v \in \mathbb{R}^s$. We say A is *definite positive* if $\langle v, Av \rangle > 0$ for all non-zero $v \in \mathbb{R}^s$. Whenever $A \in \mathrm{Sym}(s)$, the notation $A \succeq 0$

(respectively, $A \succ 0$) stands for A is positive semidefinite (respectively, positive definite).

A *semidefinite program* has the form

$$\begin{cases} p_* := \inf \sum_{i=1}^{n} c_i x_i, \\ F(x) := F_0 + x_1 F_1 + \cdots + x_n F_n \succeq 0, \\ x := (x_1, \ldots, x_n) \in \mathbb{R}^n, \end{cases} \tag{7.1}$$

where $c := (c_1, \ldots, c_n) \in \mathbb{R}^n$ and $F_i \in \mathrm{Sym}(s)$, $i = 0, \ldots, n$, for some integer s.

The *dual problem* associated to the semidefinite program (7.1) is:

$$\begin{cases} d^* := \sup -\langle F_0, Z \rangle, \\ \langle F_i, Z \rangle = c_i, \quad i = 1, \ldots, n, \\ Z \succeq 0, \ Z \in \mathrm{Sym}(s). \end{cases} \tag{7.2}$$

We can show that (7.2) is also a semidefinite program, i.e. it can be put into the same form as the primal (7.1). Furthermore, we have $d^* \le p_*$ with equality occurring when one of the following two conditions holds

(a) There exists $x \in \mathbb{R}^n$ such that $F(x) \succ 0$.
(b) There exists $Z \in \mathrm{Sym}(s)$ such that $Z \succ 0$ and $\langle F_i, Z \rangle = c_i, i = 1, \ldots, n$.

We also should mention that semidefinite programs can be solved (approximatively) in polynomial time, using the interior point methods.

7.1.2 *Sums of squares*

For $x := (x_1, \ldots, x_n) \in \mathbb{R}^n$ and $\alpha := (\alpha_1, \ldots, \alpha_n) \in \mathbb{N}^n$, let

$$x^\alpha := x_1^{\alpha_1} \ldots x_n^{\alpha_n}.$$

When $\alpha = (0, \ldots, 0)$, note that $x^\alpha = 1$. We also let $|\alpha| := \alpha_1 + \cdots + \alpha_n$ denote the total degree of the monomial x^α.

Consider a polynomial $f \in \mathbb{R}[x], f = \sum_{\alpha \in \mathbb{N}^n} f_\alpha x^\alpha$, where there are only finitely many non-zero f_α's. The *degree* of f is defined by $\deg f := \max\{|\alpha| \mid f_\alpha \neq 0\}$.

A polynomial $f \in \mathbb{R}[x]$ is called a *sum of squares* (*of polynomials*), sometimes abbreviated as "f is SOS", if it can be written as $f = \sum_{i=1}^{k} p_i^2$ for some $p_i \in \mathbb{R}[x], i = 1, \ldots, k$. Denote by $\sum \mathbb{R}[x]^2$ the set of sums of squares in $\mathbb{R}[x]$.

By definition, it is easy to see that if $f \in \mathbb{R}[x]$ is a sum of squares then $\deg f$ is even and any decomposition $f = \sum_{i=1}^{k} p_i^2$, where $p_i \in \mathbb{R}[x]$ satisfies $\deg p_i \leq \frac{\deg f}{2}$ for all i.

Given $f \in \mathbb{R}[x]$ and $S \subset \mathbb{R}^n$, the notation "$f \geq 0$ on S" means "$f(x) \geq 0$ for all $x \in S$", in this case we say that f is non-negative on S; analogously, $f > 0$ on S means that f is positive on S.

Obviously any polynomial which is a sum of squares is non-negative on $\mathbb{R}^n$. Moreover, Hilbert has shown that every non-negative polynomial of degree $2k$ in n variables is a sum of squares of polynomials if and only if we are in one of the following three cases: $(n = 1, 2k)$, $(n, 2k = 2)$, and $(n = 2, 2k = 4)$. In all other cases, in 1988, Hilbert proved the existence of a non-negative polynomial which is not sum of squares, but he gave no explicit example. The first explicit construction was found by Motzkin in 1967.

Example 7.1. Let $n = 3$ and consider the Motzkin polynomial $M(x, y, z) := z^6 + x^2 y^4 + x^4 y^2 - 3x^2 y^2 z^2 \in \mathbb{R}[x, y, z]$. By the arithmetic geometric mean inequality, we have

$$\frac{z^6 + x^2 y^4 + x^4 y^2}{3} \geq \sqrt[3]{x^6 y^6 z^6} = x^2 y^2 z^2,$$

which yields $M \geq 0$ on $\mathbb{R}^3$.

On the other hand, one can verify directly that M cannot be written as a sum of squares of polynomials. Indeed, assume to the contrary, that $M = \sum M_i^2$ for some polynomials $M_i \in \mathbb{R}[x, y, z]$. Then each M_i is homogeneous polynomial of degree 3 and hence it is some real linear combination of $x^\alpha y^\beta z^\gamma$ with $\alpha + \beta + \gamma = 3$. Looking at the coefficients of M, it is easy to see that $x^3, y^3, x^2 z, y^2 z, x z^2, y z^2$

do not appear in M_i. Thus we can write

$$M_i = a_i xy^2 + b_i x^2 y + c_i xyz + d_i z^3$$

for some scalars $a_i, b_i, c_i, d_i \in \mathbb{R}$. But then

$$M = \sum M_i^2$$
$$= \sum a_i^2 x^2 y^4 + \sum b_i^2 x^4 y^2 + \sum c_i^2 x^2 y^2 z^2 + \sum d_i^2 z^6 + \cdots ,$$

which implies that $\sum c_i^2 = -3$, a contradiction.

We next indicate how to recognize whether a polynomial can be written as a sum of squares via semidefinite programming.

Theorem 7.1. *Let* $f = \sum_{|\alpha| \leq 2k} f_\alpha x^\alpha$ *be a polynomial of degree at most $2k$ in n variables $x_1, \ldots, x_n$. The following statements are equivalent:*

(i) *f is a sum of squares.*
(ii) *There exists a matrix $Z = (Z_{\delta\beta})_{\delta,\beta} \in \mathrm{Sym}(s), Z \succeq 0$, such that*

$$\sum_{\delta+\beta=\alpha} Z_{\delta\beta} = f_\alpha \quad for \ all \ |\alpha| \leq 2k,$$

where $s := \binom{n+k}{k}$ is the binomial coefficient $n+k$ choose k.

Proof. Let $v_k := (x^\delta \, | \, |\delta| \leq k)^T \in \mathbb{R}^s$ denote the column vector containing all monomials x^δ in $\mathbb{R}[x]$ of degree $\leq k$. Then for each polynomial $p_i \in \mathbb{R}[x]$ of degree $\leq k$ we can write $p_i = \mathrm{vec}(p_i)^T v_k$, where $\mathrm{vec}(p_i)$ stands for the column vector of coefficients of p_i in the monomial basis of $\mathbb{R}[x]$, and hence

$$\sum_i p_i^2 = v_k^T \left(\sum_i \mathrm{vec}(p_i)\mathrm{vec}(p_i)^T \right) v_k.$$

Note that the matrix $\left(\sum_i \mathrm{vec}(p_i)\mathrm{vec}(p_i)^T \right) \in \mathrm{Sym}(s)$ is positive semidefinite. Furthermore, it is not hard to see that a matrix $Z \in \mathrm{Sym}(s)$ is positive semidefinite if and only if it has the form $Z = \sum_i u_i u_i^T$ for some vectors $u_i \in \mathbb{R}^s$. Therefore, $f \in \sum \mathbb{R}[x]^2$

if and only if there exists a matrix $Z \in \mathrm{Sym}(s), Z \succeq 0$, such that $f = v_k^T Z v_k$. By equating the coefficients of both polynomials f and $v_k^T Z v_k$, we get

$$\sum_{\delta+\beta=\alpha} Z_{\delta\beta} = f_\alpha, \quad |\alpha| \le 2k.$$

$\square$

More generally, given polynomials $f, h_1, \ldots, h_m \in \mathbb{R}[x]$, the problem of finding a decomposition of the form

$$f = \sigma_0 + \sigma_1 h_1 + \cdots + \sigma_m h_m$$

where $\sigma_0, \ldots, \sigma_m$ are sums of squares with a given degree bound: $\deg(\sigma_0), \deg(\sigma_j h_j) \le 2k$, can also be cast as a semidefinite program (see Proposition 7.1 below).

7.2   Representations of positive polynomials on semi-algebraic sets

The question of representing real polynomials by sum of squares of polynomials or elements from preorderings is one of the main topics in real algebra as well as real algebraic geometry. Starting with Hilbert's 17th problem of whether every non-negative real polynomial in several variables is an SOS of real rational functions, many questions have arisen in these fields and many interesting results have been found. Different kinds of Positivstellensätze and Nullstellensätze give representations of polynomials with certain properties on semi-algebraic sets.

In this section, we recall (without proof) some results representing positive polynomials on basic closed semi-algebraic sets which will be used in the next sections.

Let $h_1, \ldots, h_m \in \mathbb{R}[x]$. For simplicity, we let $h_0 := 1$ and $h^e := h_1^{e_1} \cdots h_m^{e_m}$ for $e := (e_1, \ldots, e_m) \in \{0,1\}^m$. Set

$$\mathbf{Q}(h_1, \ldots, h_m) := \left\{ \sum_{i=0}^m \sigma_i h_i \,\middle|\, \sigma_i \in \sum \mathbb{R}[x]^2 \right\},$$

$$\mathbf{P}(h_1, \ldots, h_m) := \left\{ \sum_{e \in \{0,1\}^m} \sigma_e h^e \,\middle|\, \sigma_e \in \sum \mathbb{R}[x]^2, \ e \in \{0,1\}^m \right\}.$$

Clearly, $\mathbf{Q}(h_1, \ldots, h_m) \subset \mathbf{P}(h_1, \ldots, h_m)$. We call $\mathbf{Q}(h_1, \ldots, h_m)$ (respectively, $\mathbf{P}(h_1, \ldots, h_m)$) the *quadratic module* (respectively, *preordering*) generated by $h_1, \ldots, h_m$.

Let

$$S := \{x \in \mathbb{R}^n \mid h_1(x) \geq 0, \ldots, h_m(x) \geq 0\}.$$

By definition, if $f \in \mathbf{Q}(h_1, \ldots, h_m)$ or $f \in \mathbf{P}(h_1, \ldots, h_m)$ then $f \geq 0$ on S. Conversely, we have the following certificates.

Theorem 7.2 (Krivine–Stengle's Positivstellensätz). *The following statements hold:*

(i) $f > 0$ *on S if and only if there exist polynomials $p, q \in \mathbf{P}(h_1, \ldots, h_m)$ such that $pf = 1 + q$.*

(ii) $f \geq 0$ *on S if and only if there exist an integer $k \geq 0$ and polynomials $p, q \in \mathbf{P}(h_1, \ldots, h_m)$ such that $pf = f^{2k} + q$.*

(iii) $f \equiv 0$ *on S if and only if there exists an integer $k \geq 0$ such that $-f^{2k} \in \mathbf{P}(h_1, \ldots, h_m)$.*

(iv) $S = \emptyset$ *if and only if $-1 \in \mathbf{P}(h_1, \ldots, h_m)$.*

In each case of (i)–(iii) it is clear that the "if part" gives a certificate that f is positive (non-negative, or vanishes) on S, the hard part is showing the existence of such a certificate. These certificates use polynomials in $\mathbf{P}(h_1, \ldots, h_m)$ and thus they can be checked with semidefinite programming, once a bound on the degrees has been set. However they are not directly useful for polynomial optimization problems. Indeed, in view of Theorem 7.2(i), one would need to search for the largest scalar r for which there exist $p, q \in \mathbf{P}(h_1, \ldots, h_m)$ such that $p(f - r) = 1 + q$, thus involving a term rp which cannot be dealt with directly using semidefinite programming.

To overcome this difficulty we may instead use the simpler "denominator free" positivity certificates of Schmüdgen and Putinar.

Theorem 7.3 (Schmüdgen's Positivstellensätz). *Assume that the set S is compact and $f \in \mathbb{R}[x]$. If $f > 0$ on S then $f \in \mathbf{P}(h_1, \ldots, h_m)$.*

Theorem 7.3 is an important result, but a drawback is that Schmüdgen's representation involves 2^m terms, thus leading to possibly quite large semidefinite programs. However, a major improvement is possible under some condition on the polynomials that define the semi-algebraic set S.

Definition 7.1. The quadratic module $\mathbf{Q}(h_1, \ldots, h_m)$ is said to be *Archimedean* if there exists a real number $R > 0$ such that

$$R - \sum_{i=1}^{n} x_i^2 \in \mathbf{Q}(h_1, \ldots, h_m).$$

The Archimedean condition allows easier positivity certificates using the quadratic module $\mathbf{Q}(h_1, \ldots, h_m)$.

Theorem 7.4 (Putinar's Positivstellensätz). *Assume that the quadratic module $\mathbf{Q}(h_1, \ldots, h_m)$ is Archimedean and $f \in \mathbb{R}[x]$. If $f > 0$ on S then $f \in \mathbf{Q}(h_1, \ldots, h_m)$.*

We should mention that Theorems 7.3 and 7.4 are not true if the condition "$f > 0$ on S" is replaced by "$f \geq 0$ on S", as shown by the following example.

Example 7.2. Let M be the Motzkin polynomial, i.e. $M(x, y, z) := z^6 + x^2 y^4 + x^4 y^2 - 3x^2 y^2 z^2 \in \mathbb{R}[x, y, z]$. Then M is homogeneous of degree 6, non-negative and not a sum of squares in $\mathbb{R}[x, y, z]$. In particular, $M \geq 0$ on the set $S := \{h := 1 - x^2 - y^2 - z^2 \geq 0\}$. On the other hand, it is not hard to check that if $M \in \mathbf{Q}(h)$ then M must be a sum of squares, yielding a contradiction.

Remark 7.1. (i) If $\mathbf{Q}(h_1, \ldots, h_m)$ is Archimedean then S is a compact set. In fact, by assumption, there exists a real number $R > 0$ such that $R - \sum_{i=1}^{n} x_i^2 \in \mathbf{Q}(h_1, \ldots, h_m)$. Hence $R - \sum_{i=1}^{n} x_i^2 \geq 0$ for all $x \in S$, and so S is bounded. Since S is closed, it implies that S is compact.

(ii) Assume that S is compact: there exists a constant $R > 0$ such that $S \subset \{x \in \mathbb{R}^n \mid R - \sum_{i=1}^{n} x_i^2 \geq 0\}$. Clearly, adding the quadratic constraint $R - \sum_{i=1}^{n} x_i^2 \geq 0$ in the definition of S does not change S. But with this new representation, the corresponding module $\mathbf{Q}(h_1, \ldots, h_m, R - \sum_{i=1}^{n} x_i^2)$ is Archimedean.

(iii) It should be stressed that the assumption on the compactness of S does not imply $\mathbf{Q}(h_1, \ldots, h_m)$ is Archimedean. On the other hand, it follows immediately from Theorems 7.3 that S is compact if and only if $\mathbf{P}(h_1, \ldots, h_m)$ is Archimedean.

7.3 Optimization of polynomials on compact semi-algebraic sets

Let $f, g_1, \ldots, g_l, h_1, \ldots, h_m \in \mathbb{R}[x]$ and assume that

$$S := \{x \in \mathbb{R}^n \mid g_1(x) = 0, \ldots, g_l(x) = 0, h_1(x) \geq 0, \ldots, h_m(x) \geq 0\}$$
$$\neq \emptyset.$$

Consider the optimization problem

$$f_* := \inf \{f(x) \mid x \in S\}. \tag{7.3}$$

This is an NP-hard problem even when $\deg f = 4$ and $S = \mathbb{R}^n$. A standard approach for solving (7.3) globally is Lasserre's hierarchy of semidefinite programming relaxations. It is based on a sequence of sum of squares type representations for polynomials that are non-negative on S. To describe Lasserre's hierarchy, we first introduce some notations.

In what follows, for simplicity, we write $\mathbf{Q}(\pm g_i, h_j)$ and $\mathbf{P}(\pm g_i, h_j)$ instead of the quadratic module and the preordering, respectively,

$$\mathbf{Q}(g_1, \ldots, g_l, -g_1, \ldots, -g_l, h_1, \ldots, h_m)$$

and

$$\mathbf{P}(g_1, \ldots, g_l, -g_1, \ldots, -g_l, h_1, \ldots, h_m).$$

By definition, then

$$\mathbf{Q}(\pm g_i, h_j) = \mathbf{I}(g_1, \ldots, g_l) + \mathbf{Q}(h_1, \ldots, h_m),$$
$$\mathbf{P}(\pm g_i, h_j) = \mathbf{I}(g_1, \ldots, g_l) + \mathbf{P}(h_1, \ldots, h_m),$$

where $\mathbf{I}(g_1, \ldots, g_l)$ is the ideal in the ring $\mathbb{R}[x]$ generated by the polynomials $g_1, \ldots, g_l$, that is

$$\mathbf{I}(g_1, \ldots, g_l) := \left\{ \sum_{i=1}^{l} \phi_i g_i \mid \phi_i \in \mathbb{R}[x], i = 1, \ldots, l \right\}.$$

7.3.1   Case 1: The quadratic module $\mathbf{Q}(\pm g_i, h_j)$ is Archimedean

A fundamental obstacle when one tries to find f_* is that the set S and the function $f|_S \colon S \to \mathbb{R}$ are in general far from being convex. Of course, S need not even be connected. The first step towards Lasserre's method is to convexify the problem by brute force. Obviously, we have

$$f_* = \sup\{r \in \mathbb{R} \mid f - r \geq 0 \text{ on } S\} = \sup\{r \in \mathbb{R} \mid f - r > 0 \text{ on } S\}.$$
$$(7.4)$$

In (7.4), we got rid of the usually non-convex set S by introducing a convex object which is, however, very hard to describe and not suitable for algorithmic purposes — namely, the set of polynomials non-negative (or positive) on S. The positivity certificate of Putinar (Theorem 7.4) will help us to overcome this problem. Indeed, assume $\mathbf{Q}(\pm g_i, h_j)$ is Archimedean. Then we get together with (7.4) that

$$f_* = \sup\{r \in \mathbb{R} \mid f - r \in \mathbf{Q}(\pm g_i, h_j)\}. \qquad (7.5)$$

The idea is now to relax (7.5) by introducing approximations

$$\mathbf{Q}_k(\pm g_i, h_j) := \mathbf{I}_k(g_1, \ldots, g_l) + \mathbf{Q}_k(h_1, \ldots, h_m)$$

of $\mathbf{Q}(\pm g_i, h_j)$. Here, for each integer

$$k \geq \frac{1}{2} \max\{\deg f, \deg g_1, \ldots, \deg g_l, \deg h_1, \ldots, \deg h_m\},$$

we let

$$\mathbf{I}_k(g_1, \ldots, g_l) := \left\{ \sum_{i=1}^{l} \phi_i g_i \mid \phi_i \in \mathbb{R}[x], \right.$$

$$\left. \deg(\phi_i g_i) \leq 2k, i = 1, \ldots, l \right\},$$

$$\mathbf{Q}_k(h_1, \ldots, h_m) := \left\{ \sum_{j=0}^{m} \sigma_j h_j \mid \sigma_j \in \sum \mathbb{R}[x]^2, \right.$$

$$\left. \deg(\sigma_j h_j) \leq 2k, j = 0, \ldots, m \right\},$$

where it is convenient to set $h_0 := 1$. The set $\mathbf{I}_k(g_1, \ldots, g_l)$ is called the $2k$th *truncated ideal* generated by $g_1, \ldots, g_l$, and the set $\mathbf{Q}_k(h_1, \ldots, h_m)$ is called the $2k$th *truncated quadratic module* generated by $h_1, \ldots, h_m$.

Replacing $\mathbf{Q}(\pm g_i, h_j)$ by $\mathbf{Q}_k(\pm g_i, h_j)$ in (7.5) leads to the following definition.

Definition 7.2. By *Lasserre's hierarchy* for (7.3) we mean the sequence of sum of squares relaxations:

$$\overline{f}_{*,k} := \sup\{r \in \mathbb{R} \mid f - r \in \mathbf{I}_k(g_1, \ldots, g_l) + \mathbf{Q}_k(h_1, \ldots, h_m)\}. \quad (7.6)$$

Each program (7.6) is equivalent to a semidefinite program, as shown by Proposition 7.1 below. Furthermore, the asymptotic convergence of the sequence $\{\overline{f}_{*,k}\}_k$ to the infimum f_* is guaranteed by Putinar's Positivstellensätz.

Theorem 7.5. *Assume that* $\mathbf{Q}(\pm g_i, h_j)$ *is Archimedean. Then the sequence* $\{\overline{f}_{*,k}\}_k$ *converges monotonically, increasing to* f_*.

Proof. The sequence $\overline{f}_{*,k}$ is monotonically increasing because it holds that

$$\mathbf{I}_k(g_1, \ldots, g_l) \subset \mathbf{I}_{k+1}(g_1, \ldots, g_l)$$

and

$$\mathbf{Q}_k(h_1, \ldots, h_m) \subset \mathbf{Q}_{k+1}(h_1, \ldots, h_m).$$

Let $r \in \mathbb{R}$ be such that $f - r \in \mathbf{I}_k(g_1, \ldots, g_l) + \mathbf{Q}_k(h_1, \ldots, h_m)$, i.e.

$$f - r = \sum_{i=1}^{l} \phi_i g_i + \sum_{j=0}^{m} \sigma_j h_j,$$

for some $\phi_i \in \mathbb{R}[x]$, with $\deg \phi_i \leq 2k - \deg g_i$, and $\sigma_j \in \sum \mathbb{R}[x]^2$, with $\deg \sigma_j \leq 2k - \deg h_j$. We have $f - r \geq 0$ on S. Therefore, $\overline{f}_{*,k} \leq f_*$.

On the other hand, we have $f - f_* \geq 0$ on S. Fix $\varepsilon > 0$ arbitrary. We have $f - f_* + \varepsilon > 0$ on S. By Theorem 7.4, there exist polynomials

$\phi_i, i = 1, \ldots, l$, and sums of squares $\sigma_j, j = 0, \ldots, m$, such that

$$f - f_* + \varepsilon = \sum_{i=1}^{l} \phi_i g_i + \sum_{j=0}^{m} \sigma_j h_j.$$

Hence, there exists an integer $k(\varepsilon)$ such that

$$\overline{f}_{*,k} \geq f_* - \varepsilon \quad \text{for all } k \geq k(\varepsilon).$$

Therefore, as $\varepsilon > 0$ is arbitrary, letting $\varepsilon \to 0^+$ yields the desired result. $\qquad\square$

Remark 7.2. (i) We should mention that (7.6) might not achieve its optimal value (see Remark 7.6 below). If the constraint set S has a non-empty interior, it can be shown that (7.6) attains its optimal value.

(ii) The lower bounds $\overline{f}_{*,k}$ usually have only asymptotic convergence, i.e. the finite convergence is usually not guaranteed, as shown in the following example.

Example 7.3. Consider the problem

$$f_* := \min_{(x,y,z) \in S} M(x, y, z),$$

where M is the Motzkin polynomial, i.e. $M(x, y, z) := z^6 + x^2 y^4 + x^4 y^2 - 3x^2 y^2 z^2 \in \mathbb{R}[x, y, z]$ and $S := \{(x, y, z) \in \mathbb{R}^3 \mid h(x, y, z) := 1 - x^2 - y^2 - z^2 \geq 0\}$. Then $f_* = 0$ and the origin is a global minimizer. As $Q(h)$ is Archimedean, (7.6) is feasible for k large enough. Moreover, as S has a non-empty interior, we can verify that the supremum is attained in (7.6). Hence, if $\overline{f}_{*,k} = 0$ for some k then $f \in Q(h)$, which implies that f must be a sum of squares, yielding a contradiction. Therefore, in this example, $\overline{f}_{*,k} < f_*$ for all k.

Proposition 7.1. *The problem of computing $\overline{f}_{*,k}$ is a semidefinite program.*

Proof. For simplicity we assume that $g_i \equiv 0$ for all $i = 1, \ldots, l$; the general case can be easily deduced from this case.

Let $r \in \mathbb{R}$ be such that $f - r = \sigma_0 + \sigma_1 h_1 + \cdots + \sigma_m h_m$, where $\sigma_j \in \sum \mathbb{R}[x]^2, \deg \sigma_j \leq 2k - \deg h_j, j = 0, \ldots, m$. By Theorem 7.1, for each index j there exists a symmetric square matrix $Z^{(j)} :=$ $(Z^{(j)}_{\delta\beta})_{\delta,\beta} \in \text{Sym}(s_j)$ for some integer $s_j > 0$ such that $Z^{(j)} \succeq 0$ and $\sigma_j = \sum_{\delta,\beta} Z^{(j)}_{\delta\beta} x^{\delta+\beta}$. This implies that

$$f - r = \sum_{j=0}^{m} \sigma_j h_j = \sum_{j=0}^{m} \sum_{\delta,\beta} Z^{(j)}_{\delta\beta} x^{\delta+\beta} h_j.$$

We write $h_j = \sum_\gamma h_{j\gamma} x^\gamma$. Then

$$f - r = \sum_{j=0}^{m} \sum_{\delta,\beta} \sum_{\gamma} Z^{(j)}_{\delta\beta} h_{j\gamma} x^{\delta+\beta+\gamma}.$$

Assume $f = \sum_\alpha f_\alpha x^\alpha = f_0 + \sum_{\alpha \neq 0} f_\alpha x^\alpha$. Then

$$f_0 - r + \sum_{\alpha \neq 0} f_\alpha x^\alpha = \sum_{j=0}^{m} \sum_{\delta,\beta} \sum_{\gamma} Z^{(j)}_{\delta\beta} h_{j\gamma} x^{\delta+\beta+\gamma}.$$

For each α, define the block diagonal matrix $F_\alpha := \text{diag}(F_\alpha^{(0)}, \ldots, F_\alpha^{(m)})$, where $F_\alpha^{(j)} := \left(\sum_{\delta+\beta+\gamma=\alpha} h_{j\gamma} \right)_{\delta,\beta} \in \text{Sym}(s_j)$ for $j = 0, \ldots, m$. By equating coefficients, we get

$$\begin{cases} f_0 - r = \sum_{j=0}^{m} Z^{(j)}_{00} h_{j0} = \langle F_0, Z \rangle, \\[2mm] f_\alpha = \sum_{j=0}^{m} \sum_{\delta+\beta+\gamma=\alpha} Z^{(j)}_{\delta\beta} h_{j\gamma} = \langle F_\alpha, Z \rangle, \quad \text{for } \alpha \neq 0, \end{cases}$$

where Z is the block diagonal matrix $\text{diag}(Z^{(0)}, \ldots, Z^{(m)})$. Note that $Z \succeq 0$. Hence

$$\begin{aligned} \overline{f}_{*,k} &= \sup\{r \mid f - r \in \mathbf{Q}_k(h_1, \ldots, h_m)\} \\ &= \sup\{f_0 - \langle F_0, Z \rangle \mid Z \succeq 0, f_\alpha = \langle F_\alpha, Z \rangle, \alpha \neq 0\} \\ &= f_0 + \sup\{-\langle F_0, Z \rangle \mid Z \succeq 0, f_\alpha = \langle F_\alpha, Z \rangle, \alpha \neq 0\}. \end{aligned}$$

This is the dual semidefinite program, and the proof is complete.

□

Remark 7.3. Denote by $\mathscr{X}_k$ the set of all linear maps $L \colon \mathbb{R}[x]_{2k} \to \mathbb{R}$ satisfying $L(1) = 1$ and $L \geq 0$ on $\mathbf{Q}_k(\pm g_i, h_j)$, where $\mathbb{R}[x]_{2k}$ is the vector space of all real polynomials on $\mathbb{R}^n$, of degree at most $2k$. Set

$$f_{*,k} := \inf\{L(f) \mid L \in \mathscr{X}_k\}.$$

Then it is not hard to see the following statements:

(i) The sequence $f_{*,k}$ is monotonically increasing and bounded from above by f_*.

(ii) $\overline{f}_{*,k} \leq f_{*,k} \leq f_*$. In particular, if the sequence $\{\overline{f}_{*,k}\}_k$ converges finitely to f_*, then the sequence $\{f_{*,k}\}_k$ also converges finitely to f_*.

(iii) If $\mathbf{Q}(\pm g_i, h_j)$ is Archimedean then $\lim_{k\to\infty} \overline{f}_{*,k} = \lim_{k\to\infty} f_{*,k} = f_*$.

(iv) The computation of $f_{*,k}$ as a semidefinite programming problem with computation of $\overline{f}_{*,k}$, as the dual semidefinite programming problem. Nevertheless, it is possible that the duality gap is non-zero.

We leave the verification of details to the reader as exercises.

7.3.2 *Case 2: The constraint set S is compact*

For each integer $k \geq \frac{1}{2}\max\{\deg f, \deg g_1, \dots, \deg g_l, \deg h_1, \dots, \deg h_m\}$, let

$$\mathbf{I}_k(g_1, \dots, g_l) := \left\{\sum_{i=1}^{l} \phi_i g_i \mid \phi_i \in \mathbb{R}[x], \deg(\phi_i g_i) \leq 2k\right\},$$

$$\mathbf{P}_k(h_1, \dots, h_m) := \left\{\sum_{e\in\{0,1\}^m} \sigma_e h^e \mid \sigma_e \in \sum \mathbb{R}[x]^2, \deg(\sigma_e h^e) \leq 2k\right\},$$

where, for simplicity, we put $h^e := h_1^{e_1} \dots h_m^{e_m}$ for $e := (e_1, \dots, e_m) \in \{0,1\}^m$. The set $\mathbf{P}_k(h_1, \dots, h_m)$ is called the $2k$th *truncated preordering* generated by $h_1, \dots, h_m$. Consider the sequence of sum of squares relaxations

$$\overline{f}_{*,k} := \sup\{r \in \mathbb{R} \mid f - r \in \mathbf{I}_k(g_1, \dots, g_l) + \mathbf{P}_k(h_1, \dots, h_m)\}. \quad (7.7)$$

As in the proof of Proposition 7.1, it is not hard to see that (7.7) can be reduced to a semidefinite program. Furthermore, the asymptotic convergence of the sequence $\{\overline{f}_{*,k}\}_k$ to the infimum f_* is guaranteed by Schmüdgen's Positivstellensätz.

Theorem 7.6. *Assume that S is compact. Then the sequence $\{\overline{f}_{*,k}\}_k$ converges monotonically, increasing to $f_* := \inf_{x \in S} f(x)$.*

Proof. The proof is similar to the one given in the proof of Theorem 7.5, but using Theorem 7.4 instead of Theorem 7.3. We leave the verification to the reader. $\square$

7.4 Optimality conditions and finite convergence of Lasserre's hierarchy

Let $f, g_1, \ldots, g_l, h_1, \ldots, h_m \in \mathbb{R}[x]$ and assume that

$$S := \{x \in \mathbb{R}^n \mid g_1(x) = 0, \ldots, g_l(x) = 0, h_1(x) \geq 0, \ldots, h_m(x) \geq 0\}$$
$$\neq \emptyset.$$

Consider the problem

$$f_* := \inf \{f(x) \mid x \in S\}.$$

In this section, we will show that Lasserre's hierarchy for this problem has finite convergence when the regularity, strict complementarity and second-order sufficiency conditions hold at every global minimizer, under the standard Archimedean condition.

7.4.1 *Optimality conditions*

Consider the problem

$$f_* := \inf \{f(x) \mid x \in S\}.$$

Let x^* be a local minimizer of the problem and $J(x^*) := \{j_1, \ldots, j_k\}$ be the index set of active inequality constraints at x^*. If S is regular at x^*, i.e. the gradient vectors $\nabla g_i(x^*)$, $i = 1, \ldots, l$, and $\nabla h_j(x^*)$, $j \in J(x^*)$, are linearly independent, then by Theorem 2.2 there exist (unique) Lagrange multipliers $\lambda_i, \nu_j \in \mathbb{R}$ satisfying the first-order optimality conditions

$$\nabla f(x^*) - \sum_{i=1}^{l} \lambda_i \nabla g_i(a) - \sum_{j=1}^{m} \nu_j \nabla h_j(a) = 0,$$

$$\nu_j h_j(x^*) = 0, \quad \nu_j \geq 0, \quad \text{for } j = 1, \ldots, m.$$

Recall that the *strict complementarity condition* holds at x^* if it holds that

$$\nu_1 + h_1(x^*) > 0, \ldots, \nu_m + h_m(x^*) > 0.$$

Let $L(x)$ be the associated Lagrangian function

$$L(x) := f(x) - \sum_{i=1}^{l} \lambda_i g_i(x) - \sum_{j \in J(x^*)} \nu_j h_j(x).$$

Then, under the regularity condition, the second-order necessity condition holds at x^*, i.e.

$$v^T \nabla^2 L(x^*) v \geq 0 \quad \text{for all } v \in T_{x^*} S.$$

Here $\nabla^2 L(x^*)$ is the Hessian of L at x^* and $T_{x^*}S$ stands for the *(generalized) tangent space* of S at x^*:

$$T_{x^*} S := \{v \in \mathbb{R}^n \,|\, \langle v, \nabla g_i(x^*) \rangle = 0, \; i = 1, \ldots, l, \quad \text{and}$$

$$\langle v, \nabla h_j(x^*) \rangle = 0, \; j \in J(x^*)\}.$$

If it holds that

$$v^T \nabla^2 L(x^*) v > 0 \quad \text{for all } v \in T_{x^*} S, v \neq 0,$$

we say the *second-order sufficiency condition* holds at x^*.

7.4.2 *Boundary Hessian conditions*

Let $x^* \in S$. Assume that the following conditions hold:

(i) The algebraic set

$$\mathcal{Z}(g_1, \ldots, g_l) := \{x \in \mathbb{R}^n \,|\, g_1(x) = 0, \ldots, g_l(x) = 0\}$$

is a smooth manifold of dimension $d := n - l$ in a neighborhood of x^* and there is a neighborhood U of x^* such that $\mathcal{Z}(g_1, \ldots, g_l) \cap U$ is parameterized by uniformizing parameters $t_1, t_2, \ldots, t_d$.

(ii) There exist $1 \leq j_1 < \cdots < j_k \leq m$ such that $t_1 = h_{j_1}, \ldots, t_k = h_{j_k}$ on $\mathcal{Z}(g_1, \ldots, g_l) \cap U$ and $S \cap U$ is defined by $t_1 \geq 0, \ldots, t_k \geq 0$.

The following conditions were introduced by Marshall in studying Putinar/Schmüdgen type representations for non-negative polynomials.

Definition 7.3. We say that the polynomial f satisfies the *boundary Hessian conditions* at the point $x^* \in S$ if it is expanded locally around x^* as

$$f_0 + f_1 + f_2 + \cdots,$$

with every f_i being homogeneous polynomial of degree i in $t_1, \ldots, t_d$, then the linear form

$$f_1(t_1, \ldots, t_k, 0, \ldots, 0) = c_1 t_1 + \cdots + c_k t_k$$

for some positive constants $c_1 > 0, \ldots, c_k > 0$, and the quadratic form

$$f_2(0, \ldots, 0, t_{k+1}, \ldots, t_d)$$

is positive definite in $(t_{k+1}, \ldots, t_d)$.

The following result plays an important role in the proof of Theorem 7.9 below.

Theorem 7.7 (Marshall's Nichtnegativstellensätz). *Assume that $f \geq 0$ on S. If $\mathbf{Q}(\pm g_i, h_j)$ is Archimedean and the boundary Hessian conditions hold at every zero of f on S, then there exists a polynomial $\psi \in \mathbf{Q}(h_1, \ldots, h_m)$ such that $f - \psi$ vanishes identically on the algebraic set*

$$\mathcal{Z}(g_1, \ldots, g_l) := \{x \in \mathbb{R}^n \mid g_1(x) = \cdots = g_l(x) = 0\}.$$

7.4.3 Optimality conditions and boundary Hessian conditions

The next result establishes a relationship between the optimality conditions and the boundary Hessian conditions.

Theorem 7.8. *Let $x^* \in S$ be a local minimizer of f on S. If S is regular at x^* and the strict complementarity and second-order sufficiency conditions hold at x^*, then f satisfies the boundary Hessian conditions at x^*.*

Proof. For convenience, we can generally assume $x^* = 0$, up to a shifting.

Let $J(x^*) := \{j_1, \ldots, j_k\} \subset \{1, \ldots, m\}$ be the index set of inequality constraints that are active at x^*. Since S is regular at 0, the gradient vectors

$$\nabla g_1(0), \ldots, \nabla g_l(0), \quad \nabla h_{j_1}(0), \ldots, \nabla h_{j_k}(0)$$

are linearly independent. In particular, in a neighbourhood of 0, the set $\mathcal{Z}(g_1, \ldots, g_l)$ is a manifold of dimension $d := n - l$. Up to a linear coordinate transformation, we can further assume that

$$\nabla g_1(0) = e^{d+1}, \ldots, \nabla g_l(0) = e^{d+l},$$
$$\nabla h_{j_1}(0) = e^1, \ldots, \nabla h_{j_k}(0) = e^k,$$

where $e^1, \ldots, e^n$ are the canonical basis vectors in $\mathbb{R}^n$. Note that the space $T_{x^*}S$ is determined by the vectors $e^{k+1}, \ldots, e^d$.

Define the map $\Phi \colon \mathbb{R}^n \longrightarrow \mathbb{R}^n, x \mapsto \Phi(x)$, by

$$\Phi(x) := (h_{j_1}(x), \ldots, h_{j_k}(x), x_{k+1}, \ldots, x_d, g_1(x), \ldots, g_l(x)).$$

Clearly, $\Phi(0) = 0$ and the Jacobian matrix of Φ at 0 is the identity matrix I_n. Thus, by the Implicit Function Theorem, there is a neighborhood U of 0, such that the map

$$U \to \Phi(U) \subset \mathbb{R}^n, \quad x \mapsto t := \Phi(x),$$

is a local diffeomorphism. So, $t := (t_1, \ldots, t_n)$ can serve as a coordinate system for $\mathbb{R}^n$ around 0 and $x = \Phi^{-1}(t)$.

In the t-coordinate system and in the neighborhood U, the set $\mathcal{Z}(g_1, \ldots, g_l)$ is defined by linear equations $t_{d+1} = \cdots = t_n = 0$, and $S \cap U$ defined by

$$t_1 \geq 0, \ldots, t_k \geq 0, \ t_{d+1} = \cdots = t_n = 0.$$

Let λ_i and ν_j be the Lagrange multipliers with respect to the minimizer x^*. Define the Lagrangian function

$$L(x) := f(x) - \sum_{i=1}^{l} \lambda_i g_i(x) - \sum_{r=1}^{k} \nu_{j_r} h_{j_r}(x).$$

Note that $\nabla L(0) = 0$. In the t-coordinate system, define functions

$$F(t) := f(\Phi^{-1}(t)),$$

$$\widehat{L}(t) := L(\Phi^{-1}(t)) = F(t) - \sum_{i=d+1}^{n} \lambda_{i-d} t_i - \sum_{r=1}^{k} \nu_{j_r} t_r.$$

Clearly,

$$\nabla \widehat{L}(0) = \nabla L(0) D\Phi^{-1}(0) = \nabla L(0) = 0.$$

This implies that

$$\frac{\partial F(0)}{\partial t_r} = \begin{cases} \nu_{j_r} & \text{if } r = 1, 2, \ldots, k, \\ 0 & \text{if } r = k+1, \ldots, d, \\ \lambda_{r-d} & \text{if } r = d+1, \ldots, n. \end{cases}$$

Expand F locally around 0 as

$$F(t) = f_0 + f_1(t) + f_2(t) + f_3(t) + \cdots,$$

where each f_i is a homogeneous polynomial in t and of degree i. Then we have

$$f_1(t) = \nu_{j_1} t_1 + \cdots + \nu_{j_k} t_k \quad \text{on } t_{d+1} = \cdots = t_n = 0.$$

Furthermore, for $t_{k+1}, \ldots, t_d$ near 0, it holds that

$$F(0, \ldots, 0, t_{k+1}, \ldots, t_d, 0, \ldots, 0)$$

$$= \widehat{L}(0, \ldots, 0, t_{k+1}, \ldots, t_d, 0, \ldots, 0),$$

$$= L(\Phi^{-1}(0, \ldots, 0, t_{k+1}, \ldots, t_d, 0, \ldots, 0)).$$

Let $x(t) := \Phi^{-1}(t) = (\Phi_1^{-1}(t), \dots, \Phi_n^{-1}(t))$. For all i, j, we have

$$\frac{\partial^2 \widehat{L}(t)}{\partial t_i \partial t_j} = \sum_{1 \leq r,s \leq n} \frac{\partial^2 L(x(t))}{\partial x_r \partial x_s} \cdot \frac{\partial \Phi_r^{-1}(t)}{\partial t_i} \cdot \frac{\partial \Phi_s^{-1}(t)}{\partial t_j}$$

$$+ \sum_{1 \leq r \leq n} \frac{\partial L(x(t))}{\partial x_r} \cdot \frac{\partial^2 \Phi_r^{-1}(t)}{\partial t_i \partial t_j}.$$

Note that $\nabla L(0) = 0$ and $x(0) = \Phi^{-1}(0) = 0$. Hence

$$\frac{\partial^2 \widehat{L}(0)}{\partial t_i \partial t_j} = \sum_{1 \leq r,s \leq n} \frac{\partial^2 L(0)}{\partial x_r \partial x_s} \cdot \frac{\partial \Phi_r^{-1}(0)}{\partial t_i} \cdot \frac{\partial \Phi_s^{-1}(0)}{\partial t_j}.$$

On the other hand, we have $D\Phi(0) = D\Phi^{-1}(0) = I_n$ — the identity matrix. Therefore for all $i, j = k + 1, \dots, d$,

$$\frac{\partial^2 f_2}{\partial t_i \partial t_j}\bigg|_{t=0} = \frac{\partial^2 F}{\partial t_i \partial t_j}\bigg|_{t=0} = \frac{\partial^2 \widehat{L}}{\partial t_i \partial t_j}\bigg|_{t=0} = \frac{\partial^2 L}{\partial x_i \partial x_j}\bigg|_{x=0}.$$

The strict complementarity condition implies that

$$\nu_{j_1} > 0, \dots, \nu_{j_k} > 0.$$

So, the coefficients of the linear form

$$f_1(t_1, \dots, t_k, 0, \dots, 0) = \nu_{j_1} t_1 + \cdots + \nu_{j_k} t_k$$

are all positive. Since the vector space $T_{x^*}S$ is defined by the vectors $e^{k+1}, \dots, e^d$, the second-order sufficiency condition implies that the sub-Hessian

$$\left(\frac{\partial^2 L(0)}{\partial x_i \partial x_j} \right)_{k+1 \leq i, j \leq d}$$

is positive definite. Hence, the quadratic form

$$(t_{k+1}, \dots, t_d) \mapsto f_2(0, \dots, 0, t_{k+1}, \dots, t_d, 0 \dots, 0)$$

is positive definite. Therefore, f satisfies the boundary Hessian conditions at 0. $\qquad \square$

7.4.4 *Optimality conditions and finite convergence of Lasserre's hierarchy*

For convenient we repeat the formulation of the values $\overline{f}_{*,k}$ from (7.6). For each integer $k \geq \frac{1}{2}\max\{\deg f, \deg g_1, \ldots, \deg g_l,$ $\deg h_1, \ldots, \deg h_m\}$, we put

$$\overline{f}_{*,k} := \sup\{r \in \mathbb{R} \mid f - r \in \mathbf{I}_k(g_1, \ldots, g_l) + \mathbf{Q}_k(h_1, \ldots, h_m)\},$$

where

$$\mathbf{I}_k(g_1, \ldots, g_l) := \left\{ \sum_{i=1}^{l} \phi_i g_i \mid \phi_i \in \mathbb{R}[x], \right.$$

$$\left. \deg(\phi_i g_i) \leq 2k, i = 1, \ldots, l \right\},$$

$$\mathbf{Q}_k(h_1, \ldots, h_m) := \left\{ \sum_{j=0}^{m} \sigma_j h_j \mid \sigma_j \in \sum \mathbb{R}[x]^2, \right.$$

$$\left. \deg(\sigma_j h_j) \leq 2k, j = 0, \ldots, m \right\}.$$

It was shown in Theorem 7.5 that if the quadratic module $\mathbf{Q}(\pm g_i, h_j)$ is Archimedean then the sequence $\{\overline{f}_{*,k}\}_k$ converges to the optimal value f_* of the problem $\inf_{x \in S} f(x)$. Furthermore, the convergence is finite under some generic hypotheses.

Theorem 7.9. *Assume that $\mathbf{Q}(\pm g_i, h_j)$ is Archimedean. If the regularity, strict complementarity and second-order sufficiency conditions hold at every global minimizer of the problem $\inf_{x \in S} f(x)$, then there is an integer $k_* \geq 1$ such that $\overline{f}_{*,k} = f_*$ for all $k \geq k_*$.*

In order to prove the theorem, we need the following lemma.

Lemma 7.1.

(i) *Let $\ell \geq 1$ an integer. Then, for all*

$$c \geq c_0 := \frac{1}{2\ell}\left(1 - \frac{1}{2\ell}\right)^{2\ell-1}$$

the univariate polynomial $s_c(t) := 1 + t + ct^{2\ell}$ in t is a sum of squares.

(ii) Let $\ell \geq 1$ be an integer and let $p, q \in \mathbb{R}[x]$. Then, for all $\varepsilon > 0$ and $c \in \mathbb{R}$,

$$p + \varepsilon = \theta_\varepsilon + \phi_\varepsilon,$$

where

$$\theta_\varepsilon := \varepsilon s_c\left(\frac{p}{\varepsilon}\right) + c\varepsilon^{1-2\ell}q \quad and \quad \phi_\varepsilon := -c\varepsilon^{1-2\ell}(p^{2\ell} + q).$$

(iii) In (ii), assume $c \geq c_0$ as in (i), $q \in \mathbf{Q}(h_1, \ldots, h_m)$ and $p^{2\ell} + q \in \mathbf{I}(g_1, \ldots, g_l)$. Then, there exists an integer $k > 0$ such that for all $\varepsilon > 0$,

$$\theta_\varepsilon \in \mathbf{Q}_k(h_1, \ldots, h_m) \quad and \quad \phi_\varepsilon \in \mathbf{I}_k(g_1, \ldots, g_l).$$

Proof. (i) We have $s_c'(t) = 1 + 2\ell ct^{2\ell-1} = 0$ if and only if $t = t_0 := \left(-\frac{1}{2kc}\right)^{\frac{1}{2\ell-1}}$. Furthermore, $s_c(t_0) \geq 0$ if and only if $c \geq c_0$. This implies that for all $c \geq c_0$, the polynomial s_c is non-negative on $\mathbb{R}$. Hence s_c is a sum of squares.

(ii) The proof can be done by a direct verification.

(iii) By assumption, there exist positive integers k_1 and k_2 such that

$$q \in \mathbf{Q}_{k_1}(h_1, \ldots, h_m) \quad and \quad p^{2\ell} + q \in \mathbf{I}_{k_2}(g_1, \ldots, g_l).$$

Let k_3 be an integer, $k_3 \geq \ell\frac{\deg p}{2}$. Note that $s_c\left(\frac{p}{\varepsilon}\right)$ is a sum of squares by (i) and its degree is at most $2k_3$. Hence, $\varepsilon s_c\left(\frac{p}{\varepsilon}\right) \in \mathbf{Q}_{k_3}(h_1, \ldots, h_m)$ for all $\varepsilon > 0$. Clearly, the integer number $k := \max\{k_1, k_2, k_3\}$ has the required properties. $\square$

Proof of Theorem 7.9. By Theorem 7.8, we know the boundary Hessian conditions are satisfied at every global minimizer of f on S. Then, thanks to Theorem 7.7, there exists a polynomial $\psi \in \mathbf{Q}(h_1, \ldots, h_m)$ such that $\hat{f} := f - f_* - \psi$ vanishes identically on the algebraic set $\mathcal{Z}(g_1, \ldots, g_l)$. By Real Nullstellensätz (see Theorem 7.2(iii)), there are $\ell \in \mathbb{N}$ and $q \in \sum \mathbb{R}[x]^2$ such that

$$\hat{f}^{2\ell} + q \in \mathbf{I}(g_1, \ldots, g_l).$$

Applying Lemma 7.1 to $p := \hat{f}$, q, and any $c \geq \frac{1}{2\ell}$, we get an integer k_* and polynomials $\theta_\varepsilon \in \mathbf{Q}_{k_*}(h_1, \ldots, h_m)$ and $\phi_\varepsilon \in \mathbf{I}_{k_*}(g_1, \ldots, g_l)$ such that $\hat{f} + \varepsilon = \theta_\varepsilon + \phi_\varepsilon$ for all $\varepsilon > 0$. This implies that

$$f - (f_* - \varepsilon) = \theta_\varepsilon + \psi + \phi_\varepsilon \in \mathbf{Q}_{k_*}(h_1, \ldots, h_m) + \mathbf{I}_{k_*}(g_1, \ldots, g_l).$$

Hence $\overline{f}_{*,k_*} \geq f_* - \varepsilon$. Since $\varepsilon > 0$ is arbitrary, $\overline{f}_{*,k_*} \geq f_*$. Recall that the sequence $\{\overline{f}_{*,k}\}_k$ is monotonically increasing and bounded from above by f_*. Therefore, $\overline{f}_{*,k} = f_*$ for all $k \geq k_*$.   □

Example 7.4. Let

$$f(x, y, z) := x^6 + y^6 + z^6 + 3x^2 y^2 z^2$$
$$- x^4(y^2 + z^2) - y^4(z^2 + x^2) - z^4(x^2 + y^2).$$

Consider the optimization problem:

$$f_* := \inf\{f(x, y, z) \mid g(x, y, z)$$
$$:= x^2 + y^2 + z^2 - 1 = 0, \ (x, y, z) \in \mathbb{R}^3\}.$$

The objective is the Robinson polynomial which is non-negative but not a sum of squares. The optimal value $f_* = 0$ and the global minimizers are

$$\frac{1}{\sqrt{3}}(\pm 1, \pm 1, \pm 1), \quad \frac{1}{\sqrt{2}}(\pm 1, \pm 1, 0),$$

$$\frac{1}{\sqrt{2}}(\pm 1, 0, \pm 1), \quad \frac{1}{\sqrt{2}}(0, \pm 1, \pm 1).$$

Furthermore, we have:

(a) the constraint set is regular because the unit sphere $g^{-1}(0)$ is smooth;
(b) the strict complementarity is automatically satisfied because there is no inequality constraint;
(c) the second-order sufficiency condition holds on all global minimizers. For instance, at $x^* := \frac{1}{\sqrt{3}}(1, 1, 1)$. The Lagrange multiplier with respect to the minimizer x^* is equal to zero because

$\nabla f(x^*) = 0$ and $\nabla g(x^*) = (\frac{2}{\sqrt{3}}, \frac{2}{\sqrt{3}}, \frac{2}{\sqrt{3}})^T$. Hence

$$\nabla^2 L(x^*) = \frac{4}{9} \begin{pmatrix} 2 & -1 & -1 \\ -1 & 2 & -1 \\ -1 & -1 & 2 \end{pmatrix}$$

$$= \frac{4}{9} \left(3 \begin{pmatrix} 1 & 0 & 0 \\ 0 & 1 & 0 \\ 0 & 0 & 1 \end{pmatrix} - \begin{pmatrix} 1 \\ 1 \\ 1 \end{pmatrix} \begin{pmatrix} 1 \\ 1 \\ 1 \end{pmatrix}^T \right)$$

and

$$T_{x^*} S = \{ v \in \mathbb{R}^3 \mid \langle v, (1,1,1) \rangle = 0 \}.$$

This implies that

$$v^T \nabla^2 L(x^*) v > 0 \quad \text{for all } v \in T_{x^*} S, \ v \neq 0.$$

By Theorem 7.9, the sequence $\overline{f}_{*,k}$ has finite convergence to f_*.

Remark 7.4. (i) Under the assumptions of Theorem 7.7, if in addition there is no equality constraint in the definition of S, then $f - f_* \in Q(\pm g_i, h_j)$, and so the problem of computing $\overline{f}_{*,k}$ attains its optimal value for all $k \geq k_*$.

(ii) The KKT optimality conditions are necessary for Lasserre's hierarchy to have finite convergence, as shown by the following proposition.

Proposition 7.2. *Suppose problem (7.6) achieves its optimal value. If the KKT optimality conditions fail at a global minimizer of f on S, then Lasserre's hierarchy cannot have finite convergence.*

Proof. Suppose on the contrary that $\overline{f}_{*,k} = f_*$ for some k. Since (7.6) achieves its optimum,

$$f - f_* = \sum_{i=1}^{l} \phi_i g_i + \sum_{j=0}^{m} \sigma_j h_j,$$

for some polynomials ϕ_i and sums of squares σ_j and $h_0 \equiv 1$. Let x^* be a global minimizer of the problem $\min_{x \in S} f(x)$. Note that every $g_i(x^*) = 0$ and $\sigma_j(x^*)h_j(x^*) = 0$. Differentiate the above with respect to x and evaluate it at x^*, then we get

$$\nabla f(x^*) = \sum_{i=1}^{l} \phi_i(x^*)\nabla g_i(x^*)$$

$$+ \sum_{j=0}^{m} (\sigma_j(x^*)\nabla h_j(x^*) + h_j(x^*)\nabla\sigma_j(x^*)).$$

Since every σ_j is sum of squares, $\sigma_j(x^*)h_j(x^*) = 0$ implies $h_j(x^*)\nabla\sigma_j(x^*) = 0$, and so

$$\nabla f(x^*) = \sum_{i=1}^{l} \phi_i(x^*)\nabla g_i(x^*) + \sum_{j=0}^{m} \sigma_j(x^*)\nabla h_j(x^*).$$

Therefore, the KKT optimality conditions hold at x^*, which is a contradiction. So Lasserre's hierarchy cannot have finite convergence. $\qquad\square$

In Proposition 7.2, the assumption that (7.6) achieves its optimal value cannot be dropped (this assumption is satisfied if S has non-empty interior). As a counterexample, consider the simple problem

$$\min x \quad \text{such that} \quad -x^2 \geq 0.$$

The global minimizer is 0. The KKT optimality condition fails at 0, but Lasserre's hierarchy has finite convergence: $\overline{f}_{*,k} = f_* = 0$ for all $k \geq 1$.

Remark 7.5. Assume that S is compact. Then there exists $R > 0$ such that the polynomial $R - \sum_{i=1}^{n} x_i^2$ is positive on S. By Theorem 7.3, we have $R - \sum_{i=1}^{n} x_i^2 \in \mathbf{P}(\pm g_i, h_j)$, which implies that the preordering $\mathbf{P}(\pm g_i, h_j)$ is Archimedean. Therefore, if the regularity, strict complementarity and second-order sufficiency conditions hold at every global minimizer of the problem $\inf_{x \in S} f(x)$, then the sequence $\{\overline{f}_{*,k}\}_{k \in \mathbb{N}}$ defined in (7.7) converges finitely to f_*.

Combining this fact with the results in Theorem 6.6 we have that Lasserre's hierarchy has finite convergence generically.

7.4.5 *Global optimality conditions*

In this subsection, we derive global optimality conditions for polynomial optimization which generalize the KKT optimality conditions for nonlinear optimization.

Let $(x^*, \lambda, \nu) \in S \times \mathbb{R}^l \times \mathbb{R}^m_+$ be a vector satisfying the KKT optimality conditions associated with the problem $\inf_{x \in S} f(x)$, that is

$$\nabla f(x^*) - \sum_{i=1}^{l} \lambda_i \nabla g_i(x^*) - \sum_{j=1}^{m} \nu_j \nabla h_j(x^*) = 0,$$

$$\nu_j h_j(x^*) = 0, \ \nu_j \geq 0, \quad \text{for } j = 1, \ldots, m.$$

It follows that x^* a stationary point of the Lagrangian function

$$L(x) := f(x) - \sum_{i=1}^{l} \lambda_i g_i(x) - \sum_{j=1}^{m} \nu_j h_j(x).$$

But in general, x^* is not a global minimizer of L (and may not even be a local minimizer).

On the other hand, we have the following global optimality conditions.

Theorem 7.10. *Under the assumptions of Theorem 7.9, if in addition* (7.6) *attains its optimal value, there exist polynomials $\phi_i \in \mathbb{R}[x]$ and sums of squares $\sigma_j \in \sum \mathbb{R}[x]^2$ such that*

$$f - f_* = \sigma_0 + \sum_{i=1}^{l} \phi_i g_i + \sum_{j=1}^{m} \sigma_j h_j. \tag{7.8}$$

Let $x^* \in S$ be a global minimizer of f on S. The following statements hold:

(i) $\sigma_j(x^*) h_j(x^*) = 0$ and $\sigma_j(x^*) \geq 0$ for all $j = 1, \ldots, m$.
(ii) $\nabla f(x^*) - \sum_{i=1}^{l} \phi_i(x^*) \nabla g_i(x^*) - \sum_{j=1}^{m} \sigma_j(x^*) \nabla h_j(x^*) = 0.$

(iii) x^* *is a global minimizer of the (generalized) Lagrangian function*

$$\mathscr{L}(x) := f(x) - f_* - \sum_{i=1}^{l} \phi_i(x)g_i(x) - \sum_{j=1}^{m} \sigma_j(x)h_j(x).$$

Proof. The existence of polynomials $\phi_i \in \mathbb{R}[x]$ and sums of squares $\sigma_j \in \sum \mathbb{R}[x]^2$ for which the equality (7.8) holds follows immediately from Theorem 7.9.

(i) From (7.8) and the fact that x^* is a global minimizer of f on S, we get

$$0 = f(x^*) - f_* = \sigma_0(x^*) + \sum_{i=1}^{l} \phi_i(x^*)g_i(x^*) + \sum_{j=1}^{m} \sigma_j(x^*)h_j(x^*),$$

which in turn implies (i) because $g_i(x^*) = 0$ for $i = 1,\ldots,l$, $h_j(x^*) \geq 0$ for $j = 1,\ldots,m$, and the polynomials σ_j are all SOS, hence non-negative. This also implies $\sigma_0(x^*) = 0$.

(ii) Differentiating (7.8) and using the fact that the polynomials σ_j are SOS, and using (i), yields (ii).

(iii) Since σ_0 is SOS, we have for all $x \in \mathbb{R}^n$,

$$\mathscr{L}(x) = f(x) - f_* - \sum_{i=1}^{l} \phi_i(x)g_i(x) - \sum_{j=1}^{m} \sigma_j(x)h_j(x) = \sigma_0(x) \geq 0.$$

Note that $\mathscr{L}(x^*) = \sigma_0(x^*) = 0$. Therefore, x^* is a global minimizer of $\mathscr{L}$. $\square$

The above theorem implies the following facts.

(i) The equality (7.8) should be interpreted as a *global optimality condition*.

(ii) The polynomial $\mathscr{L}(x) := f(x) - f_* - \sum_{i=1}^{l} \phi_i(x)g_i(x) - \sum_{j=1}^{m} \sigma_j(x)h_j(x)$ is a *generalized Lagrangian function*, with generalized Lagrange (polynomial) multipliers $((\phi_i),(\sigma_j)) \in (\mathbb{R}[x])^l \times (\sum \mathbb{R}[x]^2)^m$ instead of scalar multipliers $(\lambda,\nu) \in \mathbb{R}^l \times \mathbb{R}_+^m$. It is SOS (hence non-negative on $\mathbb{R}^n$), vanishes at every global minimizer $x^* \in S$, and so x^* is also a global minimizer of the generalized Lagrangian function.

(iii) The generalized Lagrange multipliers $((\phi_i),(\sigma_j)) \in (\mathbb{R}[x])^l \times (\sum \mathbb{R}[x]^2)^m$ provide a *certificate of global optimality* for $x^* \in S$ in the non-convex case exactly as the Lagrange multipliers $(\lambda, \nu) \in \mathbb{R}^l \times \mathbb{R}^m_+$ provide a certificate in the convex case.

Note from Theorem 6.6 that the global optimality condition (7.8) holds generically for polynomial optimization problems without equality constraints. (Indeed, if f is bounded from below on S then, generically, it is coercive on S. Hence, one may always add a strictly redundant constraint $M - f(x) \geq 0$ in the definition of S to ensure that S is compact.) In particular, generically, solving a polynomial optimization problem $\inf_{x \in S} f(x)$ with constraint set S and no equality constraint, reduces to solving a single semidefinite program (whose size is not known in advance).

We should also mention that in the KKT optimality conditions, only the constraints $h_j(x) \geq 0$ that are active at x^* have a possibly non-trivial associated Lagrange (scalar) multiplier ν_j. Hence the non-active constraints do not appear in the Lagrangian function L defined above. In contrast, in the global optimality condition (7.8), every constraint $h_j(x) \geq 0$ has a possibly non-trivial SOS polynomial Lagrange multiplier $\sigma_j(x)$. But if $h_j(x^*) > 0$ then necessarily $\sigma_j(x^*) = 0 = \nu_j$, as in the KKT optimality conditions.

Example 7.5. Let $n = 1$ and consider the following problem:

$$f_* := \inf\{x \mid x^2 - 1 = 0, \tfrac{1}{2} - x \geq 0\},$$

with optimal value $f_* = 1$, and global minimizer $x^* = -1$. The constraint $\tfrac{1}{2} - x \geq 0$ is not active at $x^* = -1$, but if it removed, the global minimum jumps to -1 with new global minimizer $x^* = 1$. In fact, we have the representation

$$f(x) - f_* = -(x+1) = (x + \tfrac{3}{2})(x^2 - 1) + (x+1)^2 \left(\tfrac{1}{2} - x\right),$$

which shows the important role of the constraint $\tfrac{1}{2} - x$ in the representation of $f - f_*$, via its non-trivial multiplier $x \mapsto \sigma(x) := (x+1)^2$ (which vanishes at $x^* = -1$). Note also that $\phi(x^*) = x^* + \tfrac{3}{2} = \tfrac{1}{2} = \lambda$ and $\sigma(x^*) = 0 = \nu$ are the KKT multipliers $(\lambda, \nu) \in \mathbb{R} \times \mathbb{R}_+$ in the KKT optimality conditions.

7.5   Optimization of polynomials on finite sets

Let $f, g_1, \ldots, g_l, h_1, \ldots, h_m \in \mathbb{R}[x]$ and assume that

$$S := \{x \in \mathbb{R}^n \,|\, g_1(x) = 0, \ldots, g_l(x) = 0, h_1(x) \geq 0, \ldots, h_m(x) \geq 0\}$$
$$\neq \emptyset.$$

Consider the problem

$$f_* := \inf \{f(x) \,|\, x \in S\}.$$

In this section, under the assumption that the constraint set S is finite, we show that the sequence $\{\overline{f}_{*,k}\}_{k \in \mathbb{N}}$ constructed in Section 7.3 converges finitely to f_*.

Let

$$\mathcal{Z}(g_1, \ldots, g_l) := \{x \in \mathbb{R}^n \,|\, g_1(x) = \cdots = g_l(x) = 0\}$$

is the common zero set of the polynomials $g_1, \ldots, g_l$. Then

$$S = \mathcal{Z}(g_1, \ldots, g_l) \cap \{x \in \mathbb{R}^n \,|\, h_1(x) \geq 0, \ldots, h_m(x) \geq 0\}.$$

7.5.1   *Case 1: The algebraic set $\mathcal{Z}(g_1, \ldots, g_l)$ is finite*

For convenience we repeat the formulation of the bounds $\overline{f}_{*,k}$ from (7.6). For each integer $k \geq \frac{1}{2} \max\{\deg f, \deg g_1, \ldots, \deg g_l, \deg h_1, \ldots, \deg h_m\}$, we put

$$\overline{f}_{*,k} := \sup\{r \in \mathbb{R} \,|\, f - r \in \mathbf{I}_k(g_1, \ldots, g_l) + \mathbf{Q}_k(h_1, \ldots, h_m)\},$$

where

$$\mathbf{I}_k(g_1, \ldots, g_l) := \left\{ \sum_{i=1}^{l} \phi_i g_i \,\middle|\, \phi_i \in \mathbb{R}[x], \right.$$

$$\left. \deg(\phi_i g_i) \leq 2k, i = 1, \ldots, l \right\},$$

$$\mathbf{Q}_k(h_1, \ldots, h_m) := \left\{ \sum_{j=0}^{m} \sigma_j h_j \,\middle|\, \sigma_j \in \sum \mathbb{R}[x]^2, \right.$$

$$\left. \deg(\sigma_j h_j) \leq 2k, j = 0, \ldots, m \right\}$$

with $h_0 := 1$.

It was shown in Theorem 7.5 that the sequence $\{\overline{f}_{*,k}\}_k$ converges to the optimal value f_* of the problem $\inf_{x \in S} f(x)$ provided that the quadratic module $\mathbf{Q}(\pm g_i, h_j)$ is Archimedean. Furthermore, we have the following finite convergence result.

Theorem 7.11. *If $\mathcal{Z}(g_1, \ldots, g_l)$ is finite, then there exists an integer k_* such that $\overline{f}_{*,k} = f_*$ for all $k \geq k_*$.*

Proof. If $\mathcal{Z}(g_1, \ldots, g_l) = \emptyset$, then $S = \emptyset$, and hence $f_* = +\infty$ by convention. By Krivine–Stengle's Positivstellensätz (Theorem 7.2(iv)) we have $-1 \in \sum \mathbb{R}[x]^2 + \mathbf{I}(g_1, \ldots, g_l)$, where $\mathbf{I}(g_1, \ldots, g_l)$ is the ideal generated by the polynomials $g_1, \ldots, g_l$, i.e.

$$\mathbf{I}(g_1, \ldots, g_l) := \left\{ \sum_{i=1}^{l} \phi_i g_i \;\middle|\; \phi_i \in \mathbb{R}[x] \right\}.$$

For all $r > 0$, it holds that

$$f - r = \left(1 + \frac{f}{4}\right)^2 + (-1)\left(r + \left(1 - \frac{f}{4}\right)^2\right)$$

$$\in \mathbf{I}_k(g_1, \ldots, g_l) + \mathbf{Q}_k(h_1, \ldots, h_m)$$

for all $k \gg 1$. So, $\overline{f}_{*,k} = f_*$ for all $k \gg 1$.

Now we assume that the set $\mathcal{Z}(g_1, \ldots, g_l)$ is non-empty and finite. We can write

$$\mathcal{Z}(g_1, \ldots, g_l) = \{u_1, \ldots, u_D\} \subset \mathbb{R}^n.$$

There exist the (interpolating) polynomials $\varphi_1, \ldots, \varphi_D \in \mathbb{R}[x]$ such that

$$\varphi_i(u_j) = \begin{cases} 0 & \text{if } i \neq j, \\ 1 & \text{if } i = j. \end{cases}$$

For each index i, if $f(u_i) - f_* \geq 0$, let $\psi_i := (f(u_i) - f_*)\varphi_i^2$; if $f(u_i) - f_* < 0$, then there is an index j_i such that $h_{j_i}(u_i) < 0$, and let

$$\psi_i := \left(\frac{f(u_i) - f_*}{h_{j_i}(u_i)} \right) h_{j_i} \varphi_i^2.$$

By construction, $\psi_i \in \mathbf{Q}(h_1, \ldots, h_m)$. Let $\psi := \psi_1 + \cdots + \psi_D$. We have $\psi \in \mathbf{Q}_{k_1}(h_1, \ldots, h_m)$ for some integer $k_1 > 0$. The polynomial

$$\widehat{f} := f - f_* - \psi$$

vanishes identically on $\mathcal{Z}(g_1, \ldots, g_l)$. By Theorem 7.2(iii), there exists an integer $\ell > 0$ and $q \in \sum \mathbb{R}[x]^2$ such that

$$\widehat{f}^{2\ell} + q \in \mathbf{I}(g_1, \ldots, g_l).$$

Applying Lemma 7.1 to the polynomials $p := \widehat{f}$ and q and any $c \geq \frac{1}{2\ell}$, we get an integer $k_* \geq k_1$, polynomials $\theta_\varepsilon \in \mathbf{Q}_{k_*}(h_1, \ldots, h_m)$ and $\phi_\varepsilon \in \mathbf{I}_{k_*}(g_1, \ldots, g_l)$ such that for all $\varepsilon > 0$,

$$\widehat{f} + \varepsilon = \theta_\varepsilon + \phi_\varepsilon.$$

Therefore, we get

$$f - (f_* - \varepsilon) = \sigma_\varepsilon + \phi_\varepsilon,$$

where $\sigma_\varepsilon := \theta_\varepsilon + \psi \in \mathbf{Q}_{k_*}(h_1, \ldots, h_m)$ for all $\varepsilon > 0$. This implies that $\overline{f}_{*,k_*} \geq f_*$. Note that $\overline{f}_{*,k}$ is monotonically increasing and bounded from above by f_*. So, we must have $\overline{f}_{*,k} = f_*$ for all $k \geq k_*$. $\qquad \square$

Example 7.6. A special case, which is particularly relevant to applications in combinatorial optimization, concerns the minimization of a polynomial f over the 0/1 points in a semi-algebraic set S. In other words, the equations $g_i(x) := x_i^2 - x_i = 0$ (for $i = 1, \ldots, n$) are present in the description of S; thus $\mathcal{Z}(g_1, \ldots, g_n)$ is a finite set. In this case, by Theorem 7.11, the sequence $\overline{f}_{*,k}$ converges finitely to the optimal value $f_* = \inf_{x \in S} f(x)$.

Remark 7.6. We conclude this subsection by noting that the problem of computing $\overline{f}_{*,k}$ might fail to have an optimal solution, even if the sequence $\overline{f}_{*,k}$ has finite convergence to f_*. For instance, consider the problem

$$\min_{x \in \mathbb{R}^n} x_1 \quad \text{such that } x_1^2 + \cdots + x_n^2 = 0.$$

By Theorem 7.11, we have $\overline{f}_{*,k} = 0$ for all k large enough. However, for any $\phi \in \mathbb{R}[x]$, the polynomial $\varphi := x_1 - (x_1^2 + \cdots + x_n^2)\phi$ cannot be a sum of squares because $\varphi(0) = 0, \nabla\varphi(0) \neq 0$, and 0 cannot be a minimizer of φ. This proves that the problem of computing $\overline{f}_{*,k}$ does not have a maximizer for any k large enough.

7.5.2   Case 2: The constraint set S is finite

For each integer $k \geq \frac{1}{2}\max\{\deg f, \deg g_1, \ldots, \deg g_l, \deg h_1, \ldots, \deg h_m\}$, we recall from (7.7) that the value $\overline{f}_{*,k}$ is defined by

$$\overline{f}_{*,k} := \sup\{r \in \mathbb{R} \mid f - r \in \mathbf{I}_k(g_1, \ldots, g_l) + \mathbf{P}_k(h_1, \ldots, h_m)\},$$

where

$$\mathbf{I}_k(g_1, \ldots, g_l) := \left\{\sum_{i=1}^{l} \phi_i g_i \,\middle|\, \phi_i \in \mathbb{R}[x], \deg(\phi_i g_i) \leq 2k\right\},$$

$$\mathbf{P}_k(h_1, \ldots, h_m) := \left\{\sum_{e \in \{0,1\}^m} \sigma_e h^e \,\middle|\, \sigma_e \in \sum \mathbb{R}[x]^2, \deg(\sigma_e h^e) \leq 2k\right\},$$

with $h^e := h_1^{e_1} \ldots h_m^{e_m}$ for $e := (e_1, \ldots, e_m) \in \{0,1\}^m$. We have proved in Theorem 7.6 that the sequence $\{\overline{f}_{*,k}\}_k$ converges to the optimal value f_* of the problem $\inf_{x \in S} f(x)$ provided that the set S is compact. Furthermore, we have the following finite convergence result.

Theorem 7.12. *If S is finite, then there exists an integer k_* such that $\overline{f}_{*,k} = f_*$ for all $k \geq k_*$.*

Proof.   Assume that $S = \{u_1, \ldots, u_D\} \subset \mathbb{R}^n$. Let $\varphi_1, \ldots, \varphi_D \in \mathbb{R}[x]$ be the interpolating polynomials such that $\varphi_i(u_j) = 0$ for $i \neq j$ and $\varphi_i(u_j) = 1$ for $i = j$. Since $f(u_i) - f_* \geq 0$ for all i, the polynomial

$$\psi := \sum_{i=1}^{D} (f(u_i) - f_*)\varphi_i^2$$

is a sum of squares. Furthermore, the polynomial $\widehat{f} := f - f_* - \psi$ vanishes identically on S. By Theorem 7.2(iii), there exist integers $\ell > 0$ and $k_1 > 0$ and a polynomial $q \in \mathbf{P}_{k_1}(h_1, \ldots, h_m)$ such that

$$\widehat{f}^{2\ell} + q \in \mathbf{I}_{k_1}(g_1, \ldots, g_l).$$

Applying Lemma 7.1 with $p := \widehat{f}$, q, and any $c \geq \frac{1}{2\ell}$, we get that for all $\varepsilon > 0$,

$$f - (f_* - \varepsilon) = p + \varepsilon + \psi = \sigma_\varepsilon + \phi_\varepsilon,$$

where

$$\sigma_\varepsilon := \varepsilon \left(1 + \frac{\widehat{f}}{\varepsilon} + c \left(\frac{\widehat{f}}{\varepsilon} \right)^{2\ell} \right) + c\varepsilon^{1-2\ell} q + \psi,$$

$$\phi_\varepsilon := -c\varepsilon^{1-2\ell}(\widehat{f}^{2\ell} + q) \in \mathbf{I}_{k_1}(g_1, \ldots, g_l).$$

Let $k_* \geq k_1$ be such that $\sigma_\varepsilon \in \mathbf{P}_{k_*}(h_1, \ldots, h_m)$ for all $\varepsilon > 0$. Then $f - (f_* - \varepsilon) = \sigma_\varepsilon + \phi_\varepsilon \in \mathbf{I}_k(g_1, \ldots, g_l) + \mathbf{P}_k(h_1, \ldots, h_m)$ for all $k \geq k_*$ and all $\varepsilon > 0$. Therefore, $\overline{f}_{*,k} = f_*$ for all $k \geq k_*$. $\square$

Example 7.7. Consider the optimization problem $(n = 2)$,

$$\inf \ f(x) := -x^2 - y^2$$

$$\text{subject to } h_1(x) := x^3 \geq 0, \quad h_2(x) = y^3 \geq 0,$$

$$h_3(x) = -x - y - xy \geq 0.$$

We have $S = \{(0,0)\}$ and $f_* = 0$. Furthermore, $f \equiv 0$ on S and $f^4 + q = 0$, where

$$q := \sigma_0 + \sigma_1 h_1 + \sigma_2 h_2 + \sigma_3 h_3 + \sigma_{12} h_1 h_2.$$

In the above, the sum of squares polynomials σ_i are given as:

$$\sigma_0 := (x^2 - y^2)^4 + 6(x^4 - y^4)^2,$$

$$\sigma_1 := 8 \left[x^6 \left(y + \frac{1}{2} + 4y^2 \right) + x^4 \left(\frac{x^2}{2} + 2xy + 2y^2 \right) \right.$$

$$+ x^4 \left(2y + \frac{1}{2} + 2y^2 \right) + x^2 \left(\frac{x^2}{2} + xy + 4y^2 \right)$$

$$+ 4 \left(y^4 + y^6 + y^8 \right) \Bigg],$$

$$\sigma_2 := 8 \Bigg[y^6 \left(x + \frac{1}{2} + 4x^2 \right) + y^4 \left(\frac{y^2}{2} + 2xy + 2x^2 \right)$$

$$+ y^4 \left(2x + \frac{1}{2} + 2x^2 \right) + y^2 \left(\frac{y^2}{2} + xy + 4x^2 \right)$$

$$+ 4 \left(x^4 + x^6 + x^8 \right) \Bigg],$$

$$\sigma_3 := 8[(x^8 + y^8 + x^7 + y^7 + x^6 + y^6)$$

$$+ 4x^2 y^2 (x^2 + y^2 + x^4 + y^4 + x^6 + y^6)],$$

$$\sigma_{12} := 32(x^2 + y^2 + x^4 + y^4 + x^6 + y^6).$$

Apply Lemma 7.1 with $p := f, q$, and $c = \frac{1}{4}$. For each $\varepsilon > 0$, let

$$\sigma_\varepsilon := \varepsilon \left(1 + \frac{f}{\varepsilon} + \frac{f^4}{4\varepsilon^4} \right) + \frac{1}{4\varepsilon^3} q \in \mathbf{P}_6(h_1, h_2, h_3) \quad \text{and} \quad \phi_\varepsilon := 0.$$

Then $f + \varepsilon = \phi_\varepsilon + \sigma_\varepsilon$ for all $\varepsilon > 0$. So $\overline{f}_{*,k} = 0$ for all $k \geq 6$.

7.6   Finite convergence to the unique minimizer

In this section, we will show that for almost all polynomial optimization problems, there exists a natural sequence of computationally feasible semidefinite programs, whose solutions give rise to a sequence of
points in $\mathbb{R}^n$ converging *finitely* to the optimal solution of the original
problem.

   Let $f, g_1, \ldots, g_l, h_1, \ldots, h_m \in \mathbb{R}[x]$ be polynomials of degree at
most d, and assume that

$$S := \{x \in \mathbb{R}^n \mid g_1(x) = 0, \ldots, g_l(x) = 0, h_1(x) \geq 0, \ldots, h_m(x) \geq 0\}$$

is a non-empty compact set.

For each integer $k \geq \frac{d}{2}$, we recall from (7.7) that the value $\overline{f}_{*,k}$ is defined by

$$\overline{f}_{*,k} := \sup\{r \in \mathbb{R} \mid f - r \in \mathbf{I}_k(g_1, \ldots, g_l) + \mathbf{P}_k(h_1, \ldots, h_m)\},$$

where

$$\mathbf{I}_k(g_1, \ldots, g_l) := \left\{ \sum_{i=1}^{l} \phi_i g_i \,\middle|\, \phi_i \in \mathbb{R}[x], \deg(\phi_i g_i) \leq 2k \right\},$$

$$\mathbf{P}_k(h_1, \ldots, h_m) := \left\{ \sum_{e \in \{0,1\}^m} \sigma_e h^e \,\middle|\, \sigma_e \in \sum \mathbb{R}[x]^2, \deg(\sigma_e h^e) \leq 2k \right\},$$

where we put $h^e := h_1^{e_1} \ldots h_m^{e_m}$. Denote by $\mathcal{X}_k$ the set of all linear maps $L \colon \mathbb{R}[x]_{2k} \to \mathbb{R}$ satisfying $L(1) = 1$, $L = 0$ on $\mathbf{I}_k(g_1, \ldots, g_l)$ and $L \geq 0$ on $\mathbf{P}_k(h_1, \ldots, h_m)$. Set

$$f_{*,k} := \inf\{L(f) \mid L \in \mathcal{X}_k\}. \tag{7.9}$$

Then the following statements hold:

(i) $\overline{f}_{*,k} \leq f_{*,k} \leq f_*$ for all $k \geq \frac{d}{2}$.

(ii) The sequences $\{\overline{f}_{*,k}\}_k$ and $\{f_{*,k}\}_k$ converge monotonically, increasing to the optimal value f_*.

(iii) The computation of $f_{*,k}$ as a semidefinite programming problem with computation of $\overline{f}_{*,k}$, as the dual semidefinite programming problem. Nevertheless, it is possible that the duality gap is non-zero.

Now we assume that the problem $\min_{x \in S} f(x)$ has a unique minimizer $x^* \in S$ for which the regularity, strict complementarity and second-order sufficiency conditions. (From Theorem 6.6 we know that these assumptions hold generically.) By Theorem 7.9, for all k sufficiently large, $\overline{f}_{*,k} = f_*$ and hence $f_{*,k} = f_* = f(x^*)$. Consequently, problem (7.9) attains its infimum (for example, the linear map $L \colon \mathbb{R}[x]_{2k} \to \mathbb{R}, p \mapsto p(x^*)$, is its optimal solution). Let $L_{*,k}$ be an optimal solution of problem (7.9). We need the following lemma, whose proof is omitted here.

Lemma 7.2. *For all k sufficiently large, there exist a positive integer $t \in [\frac{d}{2}, k]$, r distinct points $u_1, \ldots, u_r \in S$, and scalars $\lambda_1, \ldots, \lambda_r$ such* *that, for all $\alpha \in \mathbb{N}^n$ with $|\alpha| \leq 2t$,*

$$\lambda_1(u_1)^\alpha + \cdots + \lambda_r(u_r)^\alpha = L_{*,k}(x^\alpha),$$

$$\lambda_1 > 0, \ldots, \lambda_r > 0, \quad \lambda_1 + \cdots + \lambda_r = 1.$$

Let us write $f(x) = \sum_{\alpha \in \mathbb{N}^n} c_\alpha x^\alpha$ with $c_\alpha = 0$ for $|\alpha| > \deg f$. By applying the above lemma, we have for all k sufficiently large,

$$\lambda_1 f(u_1) + \cdots + \lambda_r f(u_r) = \sum_{|\alpha| \leq 2t} c_\alpha L_{*,k}(x^\alpha) = \sum_{|\alpha| \leq 2k} c_\alpha L_{*,k}(x^\alpha)$$

$$= L_{*,k}\left(\sum_{|\alpha| \leq 2k} c_\alpha x^\alpha\right) = L_{*,k}(f) = f_*.$$

Since every $f(u_i) \geq f_*$, each u_i must be a minimizer of f on S. Hence, $r = 1$ and $u_1 = x^*$ because x^* is the unique minimizer. Therefore, $(L_{*,k}(\pi_1), \ldots, L_{*,k}(\pi_n)) = x^*$, where $\pi_i(x) := x_i$ for $i = 1, \ldots, n$.

Summarizing, we have seen that the following is true.

Theorem 7.13. *Assume that S is compact and that the problem $\min_{x \in S} f(x)$ has a unique minimizer $x^* \in S$ for which the regularity, strict complementarity and second-order sufficiency conditions hold. Then there exists an integer number k_* such that for all $k \geq k_*$,*

$$f_{*,k} = \inf_{x \in S} f(x) \quad and \quad (L_{*,k}(\pi_1), \ldots, L_{*,k}(\pi_n)) = x^*,$$

where $f_{*,k}$ and $L_{*,k}$ are, respectively, the optimal value and some optimal solution of problem (7.9).

Bibliographic notes

Most of the materials in Section 7.3 are taken from a paper by Lasserre [76], whereas most of the materials in Sections 7.4 and 7.5 are derived from papers by Nie [112, 114]. The contents of Subsection 7.4.5 are taken from the book of Lasserre [80]. Theorem 7.13 is derived from the results in the papers by Đoạt, Hà and Phạm [32], and by Lee and Phạm [84, 85].

The field of polynomial optimization has grown considerably in the recent years. It has roots in early work of Shor [141] and later of

Nesterov [107], and the foundations were laid by the groundworks of Lasserre [75–77] and Parrilo [119, 121, 122]. The books of Lasserre [79, 80], the overviews of Laurent [82, 83] and the handbook [4] can serve as a general source about polynomial optimization.

For a survey on semidefinite programming and its multiple applications, the interested reader is referred to [146].

In his famous address at the 1900 International Congress of Mathematicians in Paris, David Hilbert proposed the question whether any positive polynomial can be written as a sum of squares of rational functions, known as Hilbert's 17th problem. This was solved by Artin in 1927 [6], a result which started the field of real algebraic geometry. For a nice discussion on historical aspects on Hilbert's 17th problem the reader is referred to [132].

The Motzkin polynomial is taken from [105]. Theorem 7.1 was discovered independently by several authors, see, for example, [21, 128].

The Positivstellensätz (Theorem 7.2) is credited to Stengle [144] but was proved earlier by Krivine [72]. Theorem 7.3 is due to Schmüdgen [137] and Theorem 7.4 is due to Putinar [130].

For more information concerning Examples 7.2 and 7.3, see [82].

The results concerning the approximation of global minimizers of polynomial optimization problems can be found in [79, 82, 138].

The boundary Hessian conditions and Theorem 7.7 are due to Marshall [98, 99].

Lemma 7.2 is a direct consequence of [113, Theorem 2.6] and [23, Theorem 1.1].

We refer the reader to the monographs [15, 99, 129] and to the overview [136] for an in-depth treatment of real algebraic aspects, and to the monograph [12] for links to convexity.

Chapter 8

Optimization of Polynomials on Non-compact Semi-algebraic Sets

Abstract

We address the problem of minimizing a polynomial f over a (not necessarily compact) basic semi-algebraic set S. The cases where the problem has optimal solution or not are considered. We will establish sums of squares certificates for non-negativity of f, respectively, on the set $\Sigma(f, S)$ of critical points of f on S and on the tangency variety $\Gamma(f, S)$. Then in each case, we construct a sequence of semidefinite programs whose optimal values converge monotonically, increasing to the optimal value of the original problem.

8.1 Unconstrained polynomial optimization I

We consider the problem of computing the global infimum of a real polynomial f on $\mathbb{R}^n$. It is well known that every global minimizer of f lies on the set $\Sigma(f)$ of critical points of f. If f attains a minimum on $\mathbb{R}^n$, it is therefore equivalent to look for the greatest lower bound of f on $\Sigma(f)$. In this section, we will prove a theorem about the existence of a sums of squares certificate for such lower bounds. Based on this certificate, we will find arbitrarily tight relaxations of the original problem that can be formulated as semidefinite programs and thus be solved efficiently.

8.1.1 *A representation of positive polynomials on semi-algebraic sets*

To begin, we introduce some new terminology.

Definition 8.1. For any polynomial $f \in \mathbb{R}[x]$ and subset $S \subset \mathbb{R}^n$, the *set of asymptotic values* of f on S, denoted by $R_\infty(f, S)$, consists of all $y \in \mathbb{R}$ for which there exists a sequence $\{x^k\}_{k \in \mathbb{N}}$ of points $x^k \in S$ such that $\lim_{k \to \infty} \|x^k\| = +\infty$ and $\lim_{k \to \infty} f(x^k) = y$.

The following result plays an important role in this chapter.

Theorem 8.1 (Schweighofer's Positivstellensätz). *Let* $f, h_1,$ $\dots, h_m \in \mathbb{R}[x]$ *and set*

$$S := \{x \in \mathbb{R}^n \mid h_1(x) \geq 0, h_2(x) \geq 0, \dots, h_m(x) \geq 0\}.$$

Suppose that

(a) f *is bounded on* S;
(b) $R_\infty(f, S)$ *is a finite subset of* $\mathbb{R}_{>0} := \{y \in \mathbb{R} \mid y > 0\}$; *and*
(c) $f > 0$ *on* S.

Then $f \in \mathbf{P}(h_1, \dots, h_m)$.

Remark that if S is compact, then $R_\infty(f, S) = \emptyset$ and hence Theorem 8.1 generalizes Theorem 7.3 (Schmüdgen's Positivstellensätz).

8.1.2 *Sets of critical points and sums of squares*

Let f be a polynomial in n real variables. We recall from Section 2.2 of Chapter 2 that *the set of critical points* of f (on $\mathbb{R}^n$) is defined by

$$\Sigma(f) := \{x \in \mathbb{R}^n \mid \nabla f(x) = 0\}.$$

Clearly, $\Sigma(f)$ is an algebraic set. Furthermore, the following statements hold:

(i) The set $K_0(f) := f(\Sigma(f))$ of critical values of f is finite (see Theorem 2.3).
(ii) If $x^* \in \mathbb{R}^n$ is a local (or global) minimizer of f then $x^* \in \Sigma(f)$.

The following result is a sums of squares certificate for non-negativity of f on $\Sigma(f)$ which is suitable for optimization purposes.

Theorem 8.2. *Let f be a polynomial in n real variables. If $f \geq 0$ on $\Sigma(f)$, then, for every $\varepsilon > 0$, there exist a sum of squares σ and polynomials $\phi_j, j = 1, \ldots, n$, such that*

$$f(x) + \varepsilon = \sigma(x) + \sum_{j=1}^{n} \phi_j(x) \frac{\partial f}{\partial x_j}(x).$$

Proof. Assume $f \geq 0$ on $\Sigma(f)$. By Theorem 2.3, the set $K_0(f) = f(\Sigma(f))$ of critical values of f (on $\mathbb{R}^n$) is finite. Hence, $f + \varepsilon$ is bounded and positive on $\Sigma(f)$, and the set

$R_\infty(f + \varepsilon, \Sigma(f))$

$\quad = \{y + \varepsilon \,|\, \text{there exists a sequence } x^k \in \Sigma(f) \text{ such that}$

$\quad\quad \|x^k\| \to +\infty \text{ and } f(x^k) \to y\}$

is a finite subset of $\mathbb{R}_{>0}$. Therefore, the polynomial $f + \varepsilon$ and the set $\Sigma(f)$ satisfy the conditions of Theorem 8.1, and the desired conclusion follows.   □

The following example of Scheiderer shows that we cannot conclude $f = \sigma + \sum_{j=1}^{n} \phi_j \frac{\partial f}{\partial x_j}$ by setting $\varepsilon = 0$ in Theorem 8.2.

Example 8.1. Consider the polynomial $f := x^8 + y^8 + z^8 + M(x, y, z) \in \mathbb{R}[x, y, z]$, where $M(x, y, z) := z^6 + x^2 y^4 + x^4 y^2 - 3x^2 y^2 z^2$ is the Motzkin polynomial. As observed earlier (see Example 7.1), M is non-negative on $\mathbb{R}^3$ but not a sum of squares. The polynomial f is non-negative over $\mathbb{R}^3$, thus over $\Sigma(f)$, but it is not a sum of squares modulo the ideal $\mathbf{I}(\nabla f)$ generated by all partial derivatives of f. Indeed, one can check that $f - \frac{M}{4} \in \mathbf{I}(\nabla f)$ and that M is not a sum of squares modulo the ideal $\mathbf{I}(\nabla f)$. We leave the proof of these claims to the reader.

8.1.3   *Application in optimization*

Let f be a polynomial in n real variables of degree d. Assume that

$$f_* := \inf\{f(x) \,|\, x \in \mathbb{R}^n\} > -\infty.$$

For each integer $k \geq d/2$, let $\overline{f}_{*,k}$ denote the optimal value of the optimization problem

$$\overline{f}_{*,k} := \sup_{r \in \mathbb{R}} r \quad \text{such that} \quad f(x) - r - \sum_{j=1}^{n} \phi_j(x) \frac{\partial f}{\partial x_j}(x) \in \sum \mathbb{R}[x]^2,$$

$$\phi_j \in \mathbb{R}[x]_{2k-d+1}, j = 1, \ldots, n.$$

Analysis similar to that in the proof of Proposition 7.1 shows that the problem of computing $\overline{f}_{*,k}$ can be reduced to a semidefinite program.

Theorem 8.3. *Assume that f attains its infimum f_*. Then the sequence $\{\overline{f}_{*,k}\}_{k \in \mathbb{N}}$ converges monotonically, increasing to f_*.*

Proof. Since $\mathbb{R}[x]_{2k-d+1}$ is a subspace of $\mathbb{R}[x]_{2(k+1)-d+1}$, the sequence $\{\overline{f}_{*,k}\}_k$ is monotonically increasing.

Let $r \in \mathbb{R}$ be such that

$$f(x) - r - \sum_{j=1}^{n} \phi_j(x) \frac{\partial f}{\partial x_j}(x) \in \sum \mathbb{R}[x]^2$$

for some polynomials $\phi_j \in \mathbb{R}[x]_{2k-d+1}$. We have

$$f(x) - r - \sum_{j=1}^{n} \phi_j(x) \frac{\partial f}{\partial x_j}(x) \geq 0 \quad \text{for all } x \in \mathbb{R}^n.$$

Let x^* be a global minimizer of f. We have $\nabla f(x^*) = 0$, and hence

$$r \leq f(x^*) - \sum_{j=1}^{n} \phi_j(x^*) \frac{\partial f}{\partial x_j}(x^*) = f(x^*) = f_*.$$

Therefore, $\overline{f}_{*,k} \leq f_*$.

On the other hand, we have $f - f_* \geq 0$ on $\Sigma(f)$. Fix $\varepsilon > 0$ arbitrary. By Theorem 8.2, there exist a sum of squares σ and polynomials $\phi_j, j = 1, \ldots, n$, such that

$$f(x) - f_* + \varepsilon = \sigma(x) + \sum_{j=1}^{n} \phi_j(x) \frac{\partial f}{\partial x_j}(x).$$

Hence, there exists an integer $k(\varepsilon)$ such that

$$\overline{f}_{*,k} \geq f_* - \varepsilon \quad \text{for all } k \geq k(\varepsilon).$$

Therefore, as $\varepsilon > 0$ is arbitrary, letting $\varepsilon \to 0^+$ yields the desired result. $\qquad\square$

Remark 8.1. (i) The condition that f attains its infimum f_* cannot be removed. A counterexample is $f(x, y) = (xy - 1)^2 + x^2$. It is easy to see that $\Sigma(f) = \{(0, 0)\}$ and $f_* = 0$. However, the sequence $\{\overline{f}_{*,k}\}$ converges monotonically, increasing to the critical value $f(0, 0) = 1 > 0 = f_*$.

(ii) If f attains its infimum on $\mathbb{R}^n$, we have only that $\lim_{k\to\infty} \overline{f}_{*,k} = f_*$. But it may be that there is no positive integer k with $\overline{f}_{*,k} = f_*$. On the other hand, this sequence has finite convergence under some assumptions, which hold generically by Theorem 5.1. Indeed, we have the next result.

Theorem 8.4. *Assume that f attains its infimum f_* and has only non-degenerate critical points. Then there exists an integer k_* such that $\overline{f}_{*,k} = f_*$ for all $k \geq k_*$.*

Proof. Our assumptions imply that $f_* = \min_{x \in \Sigma(f)} f(x)$ and that all points of $\Sigma(f)$ are isolated and hence $\Sigma(f)$ consists of finitely many points since it is an algebraic set. These facts, together with Theorem 7.11, imply the desired conclusion. $\qquad\square$

8.2   Unconstrained polynomial optimization II

Consider the problem of computing the global infimum of a real polynomial f on $\mathbb{R}^n$ which does not necessarily attain its infimum. To do this, we will replace the set $\Sigma(f)$ of critical points by the tangency variety $\Gamma(f)$. Then we establish a sums of squares certificate for non-negativity of f on its truncated tangency variety. Based on this certificate, we can find a natural sequence of semidefinite programs whose optimal values converge monotonically, increasing to the infimum of f.

8.2.1 *Truncated tangency varieties and sums of squares*

Let f be a polynomial in n real variables. Recall from Section 2.3 of Chapter 2 that the *tangency variety* of f (on $\mathbb{R}^n$) is defined by

$$\Gamma(f) := \left\{ x \in \mathbb{R}^n \,\middle|\, \text{rank} \begin{pmatrix} \dfrac{\partial f}{\partial x_1} & \dfrac{\partial f}{\partial x_2} & \cdots & \dfrac{\partial f}{\partial x_n} \\ x_1 & x_2 & \cdots & x_n \end{pmatrix} \leq 1 \right\}.$$

Define the polynomials

$$g_{ij}(x) := x_j \frac{\partial f}{\partial x_i}(x) - x_i \frac{\partial f}{\partial x_j}(x), \quad 1 \leq i < j \leq n.$$

It is clear that

$$\Gamma(f) = \{x \in \mathbb{R}^n \,|\, g_{ij}(x) = 0, \ 1 \leq i < j \leq n\},$$

which is an algebraic set. Furthermore, the following facts hold:

(i) $\Sigma(f) \subset \Gamma(f)$.
(ii) The set $T_\infty(f) = R_\infty(f, \Gamma(f))$ of tangency values (at infinity) of f (on $\mathbb{R}^n$) is finite (see Theorem 2.6).
(iii) If f is bounded from below, then the infimum value $f_* := \inf\{f(x) \,|\, x \in \mathbb{R}^n\}$ is either a critical value of f or a tangency value of f, i.e. $f_* \in K_0(f) \cup T_\infty(f)$ (see Corollary 2.1).

We shall fix a real number $M \in f(\mathbb{R}^n)$ (for instance, we can set $M := f(0)$). By the *truncated tangency variety* of f, we mean the set

$$\Gamma_M(f) := \{x \in \Gamma(f) \,|\, M - f(x) \geq 0\}.$$

By definition, then

$$\Gamma_M(f) = \{x \in \mathbb{R}^n \,|\, M - f(x) \geq 0, \ g_{ij}(x) = 0, \ 1 \leq i < j \leq n\},$$

which is a *basic closed semi-algebraic set*.

Lemma 8.1. *We have*

$$\inf\{f(x) \,|\, x \in \mathbb{R}^n\} = \inf\{f(x) \,|\, x \in \Gamma_M(f)\}.$$

Proof. It suffices to show the following inequality:

$$f_* := \inf\{f(x) \mid x \in \mathbb{R}^n\} \geq \inf\{f(x) \mid x \in \Gamma_M(f)\}.$$

We first assume that f attains its infimum f_* at some point $x^* \in \mathbb{R}^n$. We have $x^* \in \Sigma(f) \subset \Gamma(f)$, and hence $x^* \in \Gamma_M(f)$ (because $f(x^*) = f_* \leq M$). This implies the required inequality.

Now, we consider the case where f does not attain its infimum f_*. Then there is a sequence $\{x^k\}_{k \in \mathbb{N}} \subset \mathbb{R}^n$ such that $\lim_{k \to \infty} f(x^k) = f_*$. For each $k \in \mathbb{N}$, $S_k := \{x \in \mathbb{R}^n \mid \|x\|^2 = \|x^k\|^2\}$ is a non-empty compact set, and so there is $y^k \in S_k$ such that $f(y^k) = \min_{x \in S_k} f(x) \leq f(x^k)$. Then $\lim_{k \to \infty} f(y^k) = f_*$. Furthermore, by Theorem 2.1, we get $y^k \in \Gamma(f)$. This, together with the inequality $f_* < M$, implies easily that $y^k \in \Gamma_M(f)$ for all sufficiently large k, and so the desired inequality follows. $\square$

The following result is a sums of squares certificate for non-negativity of f on its truncated tangency variety which can later be read as a convergence result for a sequence of optimal values of semidefinite programs (see Theorem 8.6 below).

Theorem 8.5. *Let f be a polynomial in n real variables. Then the following conditions are equivalent:*

(i) $f \geq 0$ *on* $\mathbb{R}^n$.

(ii) $f \geq 0$ *on* $\Gamma_M(f)$.

(iii) *For every $\varepsilon > 0$, there are sums of squares σ_0 and σ_1 and polynomials $\phi_{ij}, 1 \leq i < j \leq n$, such that*

$$f(x) + \varepsilon = \sigma_0(x) + \sigma_1(x)[M - f(x)] + \sum_{1 \leq i < j \leq n} \phi_{ij}(x)g_{ij}(x).$$

Proof. (i) $\Leftrightarrow$ (ii) The implication is an immediate consequence of Lemma 8.1.

(iii) $\Rightarrow$ (ii) The implication is straightforward.

(ii) $\Rightarrow$ (iii) Assume that $f \geq 0$ on $\Gamma_M(f)$ and let ε be an arbitrary positive real number. Then the polynomial $f + \varepsilon$ is bounded and strictly positive on $\Gamma_M(f)$. On the other hand, it is easy to check that

$$R_\infty(f + \varepsilon, \Gamma_M(f)) \subset R_\infty(f + \varepsilon, \Gamma(f)) = \{y + \varepsilon \,|\, y \in T_\infty(f)\}.$$

By Theorem 2.6, $T_\infty(f)$ is a finite set. Hence, $R_\infty(f + \varepsilon, \Gamma_M(f))$ is a finite subset of $\mathbb{R}_{>0}$. Then the desired conclusion follows by applying Theorem 8.1 to the polynomial $f + \varepsilon$ and the semi-algebraic set $\Gamma_M(f)$. $\qquad\qquad\qquad\qquad\qquad\qquad\qquad\qquad\qquad\qquad\qquad\square$

8.2.2 *Application in optimization*

Let f be a polynomial in n real variables and let $f_* := \inf\{f(x) \,|\, x \in \mathbb{R}^n\}$. For each integer $k \geq \deg f/2$, we define $\overline{f}_{*,k} \in \mathbb{R} \cup \{\pm\infty\}$ as the supremum over all $r \in \mathbb{R}$ such that $f - r$ can be written as a sum

$$f(x) - r = \sigma_0(x) + \sigma_1(x)[M - f(x)] + \sum_{1 \leq i < j \leq n} \phi_{ij}(x)g_{ij}(x),$$

where σ_0, σ_1 are sums of squares and ϕ_{ij} are polynomials with $\deg \sigma_0 \leq 2k, \deg(\sigma_1(M - f)) \leq 2k$, and $\deg(\phi_{ij}g_{ij}) \leq 2k$.

As in the proof of Proposition 7.1, it is not hard to see that the problem of computing the supremum $\overline{f}_{*,k}$ can be reduced to a semidefinite program. Furthermore, we have the following general result concerning the convergence of the lower bounds.

Theorem 8.6. *The sequence $\{\overline{f}_{*,k}\}$ converges monotonically, increasing to f_*.*

Proof. In fact, if $f_* = -\infty$, then it follows from Theorem 8.5 that for every positive integer k, $\overline{f}_{*,k} = -\infty$ and there is nothing to prove. Thus, we may as well assume that $f_* > -\infty$.

According to Theorem 8.5, the number $\overline{f}_{*,k}$ is a lower bound for the infimum f_* of the polynomial f, and, by definition, this lower bound gets better as k increases

$$\cdots \leq \overline{f}_{*,k-1} \leq \overline{f}_{*,k} \leq \overline{f}_{*,k+1} \leq \cdots \leq f_*.$$

By Lemma 8.1, $f - f_* \geq 0$ on $\Gamma_M(f)$. Since $\Gamma_M(f - f_*) = \Gamma_M(f)$, it follows from Theorem 8.5 that for every $\varepsilon > 0$, there are sums of squares σ_0 and σ_1 and polynomials $\phi_{ij}, 1 \leq i < j \leq n$, such that

$$f(x) - f_* + \varepsilon = \sigma_0(x) + \sigma_1(x)[M - f(x)] + \sum_{1 \leq i < j \leq n} \phi_{ij}(x)g_{ij}(x).$$

Hence, there exists an integer $k(\varepsilon)$ such that

$$\overline{f}_{*,k} \geq f_* - \varepsilon \quad \text{for all } k \geq k(\varepsilon).$$

Since the sequence $\{\overline{f}_{*,k}\}$ is monotonically increasing and bounded from above by f_*, it follows that $\lim_{k \to \infty} \overline{f}_{*,k} = f_*$, which completes the proof of the theorem.   □

Remark 8.2. One question still unanswered is whether the truncation is necessary.

8.3   Constrained polynomial optimization I

It is possible to extend what we did in Section 8.1 to the case of constrained optimization. Indeed, let $f, g_1, \ldots, g_l, h_1, \ldots, h_m \in \mathbb{R}[x]$, and assume that the semi-algebraic set

$$S := \{x \in \mathbb{R}^n \mid g_i(x) = 0, i = 1, \ldots, l, h_j(x) \geq 0, j = 1, \ldots, m\}$$

is non-empty.

8.3.1   *Sets of critical points and sums of squares*

Recall that the *set of critical points* of f on S is defined by

$$\Sigma(f, S) := \Big\{ x \in S \mid \text{there exist } \lambda_i, \nu_j \in \mathbb{R} \text{ such that}$$

$$\nabla f(x) - \sum_{i=1}^{l} \lambda_i \nabla g_i(x) - \sum_{j=1}^{m} \nu_j \nabla h_j(x) = 0,$$

$$\nu_j h_j(x) = 0, j = 1, \ldots, m \Big\}.$$

We will show that $\Sigma(f, S)$ is a *basic closed semi-algebraic set* provided that S is regular. To do this, for each subset J of $\{1, \ldots, m\}$, define the polynomial $h_J \in \mathbb{R}[x]$ by

$$h_J(x) := \begin{cases} \prod_{j \in J} h_j(x) & \text{if } J \neq \emptyset, \\ 1 & \text{otherwise.} \end{cases}$$

If $J = \{j_1, j_2, \ldots, j_k\}$, we will denote by $p_J \in \mathbb{R}[X]$ the polynomial

$$p_J(x) := \det(A_J(x)A_J^T(x)),$$

where

$$A_J(x) := \begin{pmatrix} \dfrac{\partial f}{\partial x_1} & \dfrac{\partial f}{\partial x_2} & \cdots & \dfrac{\partial f}{\partial x_n} \\[2mm] \dfrac{\partial g_1}{\partial x_1} & \dfrac{\partial g_1}{\partial x_2} & \cdots & \dfrac{\partial g_1}{\partial x_n} \\[2mm] \vdots & \vdots & \cdots & \vdots \\[2mm] \dfrac{\partial g_l}{\partial x_1} & \dfrac{\partial g_l}{\partial x_2} & \cdots & \dfrac{\partial g_l}{\partial x_n} \\[2mm] \dfrac{\partial h_{j_1}}{\partial x_1} & \dfrac{\partial h_{j_1}}{\partial x_2} & \cdots & \dfrac{\partial h_{j_1}}{\partial x_n} \\[2mm] \vdots & \vdots & \cdots & \vdots \\[2mm] \dfrac{\partial h_{j_k}}{\partial x_1} & \dfrac{\partial h_{j_k}}{\partial x_2} & \cdots & \dfrac{\partial h_{j_k}}{\partial x_n} \end{pmatrix}$$

is the $(l+k+1) \times n$ matrix. Observe that $p_J(x) = 0$ if and only if the vectors $\nabla f(x)$, $\nabla g_i(x), i = 1, \ldots, l$, and $\nabla h_j(x), j \in J$, are linearly dependent.

Lemma 8.2. *If the set S is regular, then*

$$\Sigma(f, S) = \{x \in S \,|\, h_J(x)p_{J^c}(x) = 0 \text{ for all } J \subset \{1, \ldots, m\}\},$$

where and in the following, we use the notation $J^c := \{1, \ldots, m\} \setminus J$. In particular, $\Sigma(f, S)$ is a basic closed semi-algebraic set.

Proof. Let $x \in \Sigma(f, S)$. Take an arbitrary subset J of the set $\{1, \ldots, m\}$. We claim that $h_J(x)p_{J^c}(x) = 0$. Indeed, by definition, there exist real numbers λ_i, ν_j such that

$$\nabla f(x) - \sum_{i=1}^{l} \lambda_i \nabla g_i(x) - \sum_{j=1}^{m} \nu_j \nabla h_j(x) = 0, \quad \text{and}$$

$$\nu_j h_j(x) = 0, \ j = 1, \ldots, m.$$

It follows that

$$\nabla f(x) - \sum_{i=1}^{l} \lambda_i \nabla g_i(x) - \sum_{j \in J(x)} \nu_j \nabla h_j(x) = 0.$$

(Recall that $J(x) = \{j \mid h_j(x) = 0\}$.) Consequently, the gradient vectors $\nabla f(x), \nabla g_i(x), i = 1, \ldots, l$, and $\nabla h_j(x), j \in J(x)$, are linearly dependent, which is equivalent to the fact that $p_{J(x)}(x) = 0$. This implies that $p_{J^c}(x) = 0$ provided that $J \cap J(x) = \emptyset$. On the other hand, if $J \cap J(x) \neq \emptyset$, then $h_J(x) = 0$. Therefore, $h_J(x)p_{J^c}(x) = 0$.

Conversely, let $x \in S$ be such that $h_J(x)p_{J^c}(x) = 0$ for all $J \subset \{1, \ldots, m\}$. By definition, $h_{J(x)^c}(x) > 0$. Then $p_{J(x)}(x) = 0$, and so the vectors $\nabla f(x), \nabla g_i(x), i = 1, \ldots, l$, and $\nabla h_j(x), j \in J(x)$, are linearly dependent. This implies that there exist real numbers $\kappa, \lambda_i, \nu_j, j \in J(x)$, at least one of which is different from zero, such that

$$\kappa \nabla f(x) - \sum_{i=1}^{l} \lambda_i \nabla g_i(x) - \sum_{j \in J(x)} \nu_j \nabla h_j(x) = 0.$$

Since S is regular, we can assume that $\kappa = 1$. For each $j \notin J(x)$, we let $\nu_j = 0$. Then

$$\nabla f(x) - \sum_{i=1}^{l} \lambda_i \nabla g_i(x) - \sum_{j=1}^{m} \nu_j \nabla h_j(x) = 0, \quad \text{and}$$

$$\nu_j h_j(x) = 0, \ j = 1, \ldots, m.$$

Therefore, $x \in \Sigma(f, S)$, which completes the proof of the lemma. $\square$

Recall that

$$\mathbf{P}(h_1, \ldots, h_m) := \left\{ \sum_{e \in \{0,1\}^m} \sigma_e h_1^{e_1} \ldots h_m^{e_m} \, \middle| \, \sigma_e \in \sum \mathbb{R}[x]^2 \right\}$$

is the *preordering* generated by the polynomials $h_1, \ldots, h_m$. We will denote by $\mathbf{I}(f, S)$ the *ideal generated by the polynomials* g_i and $h_J p_{J^c}$, i.e.

$$\mathbf{I}(f, S) := \left\{ \sum_{i=1}^{l} \phi_i g_i + \sum_{J \subset \{1, \ldots, m\}} \psi_J h_J p_{J^c} \,\middle|\, \phi_i, \psi_J \in \mathbb{R}[x] \right\}.$$

We are now able to state a sums of squares certificate for non-negativity of f on the set $\Sigma(f, S)$ which is suitable for optimization purposes.

Theorem 8.7. *Assume S is regular. If $f \geq 0$ on $\Sigma(f, S)$, then we have for all $\varepsilon > 0$,*

$$f + \varepsilon \in \mathbf{I}(f, S) + \mathbf{P}(h_1, \ldots, h_m).$$

Proof. Assume $f \geq 0$ on $\Sigma(f, S)$ and let $\varepsilon > 0$ fixed arbitrary. By Theorem 2.3, the set $K_0(f, S) = f(\Sigma(f, S))$ of critical values of f on S is finite. Hence, the polynomial $f + \varepsilon$ is bounded and strictly positive on $\Sigma(f, S)$, and the set

$$R_\infty(f + \varepsilon, \Sigma(f, S))$$
$$= \{y + \varepsilon \,|\, \text{there exists a sequence } \{x^k\} \subset \Sigma(f, S)$$
$$\text{such that } \|x^k\| \to +\infty \text{ and } f(x^k) \to y\}$$

is a finite subset of $\{y \in \mathbb{R} \,|\, y > 0\}$. Then the desired conclusion follows by applying Theorem 8.1 to the polynomial $f + \varepsilon$ and the basic closed semi-algebraic set $\Sigma(f, S)$. $\qquad\qquad\square$

8.3.2 *Application in optimization*

Assume that f is bounded from below on S. Consider the problem

$$f_* := \inf_{x \in S} f(x) > -\infty.$$

We now construct a sequence of semidefinite programs whose optimal values converge monotonically, increasing to the value f_*.

To do this, for each integer $k \geq \frac{1}{2}\max\{\deg f, \deg g_1, \ldots, \deg g_l, \deg h_1, \ldots, \deg h_m\}$, we put

$$
\mathbf{I}_k(f, S) := \left\{ \sum_{i=1}^{l} \phi_i g_i + \sum_{J \subset \{1,\ldots,m\}} \psi_J h_J p_{J^c} \,\middle|\, \phi_i, \psi_J \in \mathbb{R}[x], \right.
$$

$$
\left. \deg(\phi_i g_i), \deg(\psi_J h_J p_{J^c}) \leq 2k \right\},
$$

$$
\mathbf{P}_k(h_1, \ldots, h_m) := \left\{ \sum_{e \in \{0,1\}^m} \sigma_e h_1^{e_1} \ldots h_m^{e_m} \,\middle|\, \sigma_e \in \sum \mathbb{R}[x]^2, \right.
$$

$$
\left. \deg(\sigma_e h_1^{e_1} \ldots h_m^{e_m}) \leq 2k \right\}.
$$

Let $\overline{f}_{*,k}$ denote the optimal value of the optimization problem

$$
\overline{f}_{*,k} := \sup_{r \in \mathbb{R}} r \quad \text{such that } f - r \in \mathbf{I}_k(f, S) + \mathbf{P}_k(h_1, \ldots, h_m).
$$

Again, the argument used in the proof of Proposition 7.1 shows that the computation of $\overline{f}_{*,k}$ amounts to solving a semidefinite program for each fixed k.

We conclude this section by interpreting Theorem 8.7 as a convergence result concerning the optimal values $\overline{f}_{*,k}$ of the proposed relaxations.

Theorem 8.8. *Assume S is regular. If f attains its infimum f_* over S, then the sequence $\{\overline{f}_{*,k}\}$ converges monotonically, increasing to f_*.*

Proof. This is similar to Theorem 8.3 and is left as an exercise.

$\square$

Remark 8.3. The assumption in Theorem 8.8 that f has a minimum at some point cannot be removed, as shown in the following example.

Example 8.2. Let $f(x, y, z) := (xy-1)^2 + x^2 + z$ and $S := \{(x, y, z) \in \mathbb{R}^3 \mid g(x, y, z) := z = 0\}$. We have $f_* := \inf_{(x,y,z) \in S} f(x, y, z) = 0$ and $\Sigma(f, S) = \{(0, 0, 0)\}$. However, the sequence $\{\overline{f}_{*,k}\}$ converges monotonically, increasing to the critical value $f(0, 0, 0) = 1 > 0 = f_*$.

Finally, we should mention that, by Theorem 7.12, $\overline{f}_{*,k} = f_*$ for all k large enough provided that the following two properties are satisfied: (a) the problem $\min_{x \in S} f(x)$ has a global minimizer and (b) the set $\Sigma(f, S)$ is finite. On the other hand, these properties hold generically as shown in Theorem 6.6. Therefore, the sequence $\{\overline{f}_{*,k}\}_k$ converges finitely to f_* generically.

8.4 Constrained polynomial optimization II

Let $f, g_1, \ldots, g_l, h_1, \ldots, h_m \in \mathbb{R}[x]$, and assume that the semi-algebraic set

$$S := \{x \in \mathbb{R}^n \mid g_i(x) = 0, i = 1, \ldots, l, h_j(x) \geq 0, j = 1, \ldots, m\}$$

is non-empty. In this section, we show a sums of squares certificate for non-negativity of f on its truncated tangency variety which is suitable for optimization purposes.

8.4.1 *Truncated tangency varieties and sums of squares*

Recall from Section 2.3 of Chapter 2 that the *tangency variety of f on S* is defined as follows:

$$\Gamma(f, S) := \left\{ x \in S \middle| \text{there exist } \kappa, \lambda_i, \nu_j, \mu \in \mathbb{R}, \text{ not all zero, such that} \right.$$

$$\kappa \nabla f(x) - \sum_{i=1}^{l} \lambda_i \nabla g_i(x) - \sum_{j=1}^{m} \nu_j \nabla h_j(x) - \mu x = 0,$$

$$\left. \nu_j h_j(x) = 0, j = 1, \ldots, m \right\}.$$

We will show that $\Gamma(f, S)$ is a basic closed semi-algebraic set. To do this, for each subset J of $\{1, \ldots, m\}$, consider the polynomial

$$h_J(x) := \begin{cases} \displaystyle\prod_{j \in J} h_j(x) & \text{if } J \neq \emptyset, \\ 1 & \text{otherwise.} \end{cases}$$

If $J = \{j_1, j_2, \ldots, j_k\}$, we will denote by $p_J \in \mathbb{R}[X]$ the following polynomial:

$$p_J(x) := \det(A_J(x) A_J^T(x)),$$

where

$$A_J(x) := \begin{pmatrix} \dfrac{\partial f}{\partial x_1} & \dfrac{\partial f}{\partial x_2} & \cdots & \dfrac{\partial f}{\partial x_n} \\[2mm] \dfrac{\partial g_1}{\partial x_1} & \dfrac{\partial g_1}{\partial x_2} & \cdots & \dfrac{\partial g_1}{\partial x_n} \\[2mm] \vdots & \vdots & \cdots & \vdots \\[2mm] \dfrac{\partial g_l}{\partial x_1} & \dfrac{\partial g_l}{\partial x_2} & \cdots & \dfrac{\partial g_l}{\partial x_n} \\[2mm] \dfrac{\partial h_{j_1}}{\partial x_1} & \dfrac{\partial h_{j_1}}{\partial x_2} & \cdots & \dfrac{\partial h_{j_1}}{\partial x_n} \\[2mm] \vdots & \vdots & \cdots & \vdots \\[2mm] \dfrac{\partial h_{j_k}}{\partial x_1} & \dfrac{\partial h_{j_k}}{\partial x_2} & \cdots & \dfrac{\partial h_{j_k}}{\partial x_n} \\[2mm] x_1 & x_2 & \cdots & x_n \end{pmatrix}$$

is the $(l + k + 2) \times n$-matrix. Observe that $p_J(x) = 0$ if and only if the vectors $\nabla f(x), \nabla g_i(x), i = 1, \ldots, l$, and $\nabla h_j(x), j \in J$, and the vector x are linearly dependent.

Lemma 8.3. *We have*

$$\Gamma(f, S) = \{x \in S \mid h_J(x) p_{J^c}(x) = 0 \text{ for all } J \subset \{1, \ldots, m\}\},$$

where $J^c := \{1, \ldots, m\} \setminus J$. *In particular,* $\Gamma(f, S)$ *is a basic closed semi-algebraic set.*

Proof. Let $x \in \Gamma(f, S)$. Take an arbitrary subset J of the set $\{1, \ldots, m\}$. We claim that $h_J(x)p_{J^c}(x) = 0$. Indeed, by definition, there exist real numbers $\kappa, \lambda_i, \nu_j, \mu$, at least one of which is different from zero, such that

$$\kappa \nabla f(x) - \sum_{i=1}^{l} \lambda_i \nabla g_i(x) - \sum_{j=1}^{m} \nu_j \nabla h_j(x) - \mu x = 0,$$

$$\nu_j h_j(x) = 0, \quad j = 1, \ldots, m.$$

It follows that

$$\kappa \nabla f(x) - \sum_{i=1}^{l} \lambda_i \nabla g_i(x) - \sum_{j \in J(x)} \nu_j \nabla h_j(x) - \mu x = 0.$$

Consequently, the vectors $\nabla f(x), \nabla g_i(x), i = 1, \ldots, l$, and $\nabla h_j(x)$, $j \in J(x)$, and the vector x are linearly dependent, which is equivalent to the fact that $p_{J(x)}(x) = 0$. This implies that $p_{J^c}(x) = 0$ provided that $J \cap J(x) = \emptyset$. On the other hand, if $J \cap J(x) \neq \emptyset$, then $h_J(x) = 0$. Therefore, $h_J(x)p_{J^c}(x) = 0$.

Conversely, let $x \in S$ be such that $h_J(x)p_{J^c}(x) = 0$ for all $J \subset \{1, \ldots, m\}$. By definition, $h_{J(x)^c}(x) > 0$. Then $p_{J(x)}(x) = 0$, and so the vectors $\nabla f(x), \nabla g_i(x), i = 1, \ldots, l$, and $\nabla h_j(x), j \in J(x)$, and the vector x are linearly dependent. This implies that there exist real numbers $\kappa, \lambda_i, \nu_j, j \in J(x), \mu$, at least one of which is different from zero, such that

$$\kappa \nabla f(x) - \sum_{i=1}^{l} \lambda_i \nabla g_i(x) - \sum_{j \in J(x)} \nu_j \nabla h_j(x) - \mu x = 0.$$

For each $j \notin J(x)$, we let $\nu_j = 0$. Then

$$\kappa \nabla f(x) - \sum_{i=1}^{l} \lambda_i \nabla g_i(x) - \sum_{j=1}^{m} \nu_j \nabla h_j(x) - \mu x = 0,$$

$$\nu_j h_j(x) = 0, j = 1, \ldots, m.$$

Therefore, $x \in \Gamma(f, S)$. This completes the proof of the lemma. $\square$

Let M be any real number such that $M \geq f(x^0)$ for some $x^0 \in S$. By the *truncated tangency variety* of f on S, we mean the set

$$\Gamma_M(f, S) := \{x \in \Gamma(f, S) \,|\, M - f(x) \geq 0\}.$$

We can write

$$\Gamma_M(f, S) = \{x \in S \,|\, M - f(x) \geq 0 \text{ and }$$
$$h_J(x)p_{J^c}(x) = 0 \text{ for all } J \subset \{1, \ldots, m\}\},$$

and so $\Gamma_M(f, S)$ is a basic closed semi-algebraic set.

Lemma 8.4. *We have*

$$\inf\{f(x) \,|\, x \in S\} = \inf\{f(x) \,|\, x \in \Gamma_M(f, S)\}.$$

Proof. The proof is similar to Theorem 2.4. Indeed, it suffices to show the following inequality:

$$f_* := \inf\{f(x) \,|\, x \in S\} \geq \inf\{f(x) \,|\, x \in \Gamma_M(f, S)\}.$$

To do this, we first assume that f attains its infimum f_* on S at some point $x^* \in S$. Then it follows from the Fritz-John optimality conditions that $x^* \in \Gamma(f, S)$, and so $x^* \in \Gamma_M(f, S)$ because $f(x^*) = f_* \leq M$. This shows the required inequality.

We now assume that f does not attain its infimum f_* on S. Then there is a sequence $\{x^k\}_{k \in \mathbb{N}} \subset S$ such that $\lim_{k \to \infty} f(x^k) = f_*$. For each $k \in \mathbb{N}$, since $S_k := \{x \in S \,|\, \|x\|^2 = \|x^k\|^2\}$ is a non-empty compact set, there is $y^k \in S_k$ such that $f(y^k) = \min_{x \in S_k} f(x) \leq f(x^k)$. We have $\lim_{k \to \infty} f(y^k) = f_*$. Furthermore, by the Fritz-John optimality conditions, $y^k \in \Gamma(f, S)$. This, together with the inequality $f_* < M$, implies easily that $y^k \in \Gamma_M(f, S)$ for k sufficiently large, and so the desired inequality follows. $\quad\square$

Let $h_{m+1} := M - f \in \mathbb{R}[x]$. Recall that the *preordering generated* by the polynomials $h_1, \ldots, h_{m+1}$ is defined by

$$\mathbf{P}(h_1, \ldots, h_{m+1}) := \left\{ \sum_{e \in \{0,1\}^{m+1}} \sigma_1 h_1^{e_1} \ldots h_{m+1}^{e_{m+1}} \,\middle|\, \sigma_e \in \sum \mathbb{R}[x]^2 \right\}.$$

We will denote by $\mathbf{I}(f, S)$ the *ideal generated by the polynomials* g_i and $h_{J} p_{J^c}$, i.e.

$$\mathbf{I}(f, S) := \left\{ \sum_{i=1}^{l} \phi_i g_i + \sum_{J \subset \{1,\ldots,m\}} \psi_J h_J p_{J^c} \,\middle|\, \phi_i, \psi_J \in \mathbb{R}[x] \right\}.$$

Here comes one of the main results of this section which is interesting on its own but can later be read as a convergence result for a sequence of optimal values of semidefinite programs (see Theorem 8.10 below).

Theorem 8.9. *Assume the set S is regular. Then the following conditions are equivalent:*

(i) $f \geq 0$ *on* S.
(ii) $f \geq 0$ *on* $\Gamma_M(f, S)$.
(iii) *We have for all* $\varepsilon > 0$,

$$f + \varepsilon \in \mathbf{I}(f, S) + \mathbf{P}(h_1, \ldots, h_{m+1}).$$

Proof. (i) $\Leftrightarrow$ (ii) This is an immediate consequence of Lemma 8.4.

(iii) $\Rightarrow$ (ii) The implication is straightforward.

(ii) $\Rightarrow$ (iii) Assume that $f \geq 0$ on $\Gamma_M(f, S)$ and let ε be any positive real number. Then the polynomial $f + \varepsilon$ is bounded and strictly positive on the truncated tangency variety $\Gamma_M(f, S)$. On the other hand, we have

$$R_\infty(f + \varepsilon, \Gamma_M(f, S)) \subset \{y + \varepsilon \,|\, y \in R_\infty(f, \Gamma(f, S))\}$$
$$= \{y + \varepsilon \,|\, y \in T_\infty(f, S)\}.$$

By Theorem 2.6, $T_\infty(f, S)$ is a finite set. Hence, $R_\infty(f + \varepsilon, \Gamma_M(f, S))$ is a finite subset of $\mathbb{R}_{>0}$. Then the statement follows by applying Theorem 8.1 to the polynomial $f + \varepsilon$ and the semi-algebraic set $\Gamma_M(f, S)$. $\qquad\square$

8.4.2 Application in optimization

Let $f, g_1, \ldots, g_l, h_1, \ldots, h_m \in \mathbb{R}[x]$, and assume that the semi-algebraic set

$$S := \{x \in \mathbb{R}^n \,|\, g_i(x) = 0, i = 1, \ldots, l, h_j(x) \geq 0, j = 1, \ldots, m\}$$

is non-empty. Let

$$f_* := \inf_{x \in S} f(x).$$

We will construct a sequence of semidefinite programs whose optimal values converge monotonically, increasing to the value f_*. To do this, let $M \geq f(x^0)$ for some $x^0 \in S$. For each integer $k \geq \frac{1}{2} \max\{\deg f, \deg g_1, \ldots, \deg g_l, \deg h_1, \ldots, \deg h_m\}$, we put

$$\mathbf{I}_k(f, S) := \left\{ \sum_{i=1}^{l} \phi_i g_i + \sum_{J \subset \{1, \ldots, m\}} \psi_J h_J p_{J^c} \,\middle|\, \phi_i, \psi_J \in \mathbb{R}[x], \right.$$

$$\left. \deg(\phi_i g_i), \deg(\psi_J h_J p_{J^c}) \leq 2k \right\},$$

$$\mathbf{P}_k(h_1, \ldots, h_{m+1}) := \left\{ \sum_{e \in \{0,1\}^{m+1}} \sigma_e h_1^{e_1} \ldots h_{m+1}^{e_{m+1}} \,\middle|\, \sigma_e \in \sum \mathbb{R}[x]^2, \right.$$

$$\left. \deg(\sigma_e h_1^{e_1} \ldots h_{m+1}^{e_{m+1}}) \leq 2k \right\},$$

where $h_{m+1} := M - f \in \mathbb{R}[x]$. Let $\overline{f}_{*,k}$ denote the optimal value of the optimization problem

$$\overline{f}_{*,k} := \sup_{r \in \mathbb{R}} r$$

such that $f - r \in \mathbf{I}_k(f, S) + \mathbf{P}_k(h_1, \ldots, h_{m+1})$.

As in the proof of Proposition 7.1, we can show that this problem can be reduced to a semidefinite program.

Theorem 8.10. *If the set S is regular, then the sequence $\{f_{*,k}\}$ converges monotonically, increasing to f_*.*

Proof. In fact, if $f_* = -\infty$, then it is easily seen from Theorem 8.9 that $f_{*,k} = -\infty$ for every positive integer k, and there is nothing to prove. Thus, we may as well assume that $f_* > -\infty$.

According to Theorem 8.9, the number $\overline{f}_{*,k}$ is a lower bound for the infimum f_* of the polynomial f, and, by definition, this lower bound gets better as k increases

$$\cdots \leq \overline{f}_{*,k-1} \leq \overline{f}_{*,k} \leq \overline{f}_{*,k+1} \leq \cdots \leq f_*.$$

By Lemma 8.4, $f - f_* \geq 0$ on $\Gamma_M(f, S)$. Let ε be any positive constant. Since $\Gamma_M(f - f_*, S) = \Gamma_M(f, S)$, it follows from Theorem 8.9 that

$$f - f_* + \varepsilon \in \mathbf{I}(f, S) + \mathbf{P}(h_1, \ldots, h_{m+1}).$$

Hence, there exists an integer $k(\varepsilon)$ such that

$$\overline{f}_{*,k} \geq f_* - \varepsilon \quad \text{for all } k \geq k(\varepsilon).$$

Since the sequence $\{\overline{f}_{*,k}\}$ is monotonically increasing and bounded from above by f_*, it follows that $\lim_{k \to \infty} \overline{f}_{*,k} = f_*$, which completes the proof of the theorem. $\qquad\square$

8.5 Further remarks

Let $g_1, \ldots, g_l, h_1, \ldots, h_m \in \mathbb{R}[x]$ be polynomials and assume that the set

$$S := \{x \in \mathbb{R}^n \mid g_1(x) = 0, \ldots, g_l(x) = 0, h_1(x) \geq 0, \ldots, h_m(x) \geq 0\}$$

is non-empty. We will show that for almost all polynomial $f \in \mathbb{R}[x]$, there exists a natural sequence of computationally feasible semidefinite programs, whose solutions give rise to a sequence of points in $\mathbb{R}^n$ converging *finitely* to the optimal solution of the problem $\inf_{x \in S} f(x)$.

Indeed, it was shown in Theorem 6.6 that, for a generic polynomial f, which is bounded from below on S, we have

(i) The restriction of f on S is coercive.
(ii) The optimal problem $\inf_{x \in S} f(x)$ has a unique minimizer $x^* \in S$ for which the strict complementarity and second-order sufficiency conditions hold.

Let M be any real number such that $M > f(x^0)$ for some $x^0 \in S$. Then the basic closed semi-algebraic set

$$\begin{aligned} S_M := \{x \in S \mid M - f(x) \geq 0\} \\ = \{x \in \mathbb{R}^n \mid g_1(x) = 0, \ldots, g_l(x) = 0, \\ h_1(x) \geq 0, \ldots, h_{m+1}(x) \geq 0\} \end{aligned}$$

is non-empty compact, where $h_{m+1} := M - f \in \mathbb{R}[x]$. Moreover, it is clear that $x^* \in S_M$ and

$$f(x^*) = \inf_{x \in S} f(x) = \min_{x \in S_M} f(x).$$

For each integer

$$k \geq \frac{1}{2} \max\{\deg f, \deg g_1, \ldots, \deg g_l, \deg h_1, \ldots, \deg h_m\},$$

let $\mathcal{X}_k$ be the set of all linear maps $L \colon \mathbb{R}[x]_{2k} \to \mathbb{R}$ satisfying $L(1) = 1$, $L = 0$ on $\mathbf{I}_k(g_1, \ldots, g_l)$ and $L \geq 0$ on $\mathbf{P}_k(h_1, \ldots, h_{m+1})$, where

$$\mathbf{I}_k(g_1, \ldots, g_l) := \left\{ \sum_{i=1}^{l} \phi_i g_i \,\middle|\, \phi_i \in \mathbb{R}[x], \deg(\phi_i g_i) \leq 2k \right\},$$

$$\mathbf{P}_k(h_1, \ldots, h_{m+1}) := \left\{ \sum_{e \in \{0,1\}^{m+1}} \sigma_e h_1^{e_1} \ldots h_{m+1}^{e_{m+1}} \,\middle|\, \sigma_e \in \sum \mathbb{R}[x]^2, \right.$$

$$\left. \deg(\sigma_e h_1^{e_1} \ldots h_{m+1}^{e_{m+1}}) \leq 2k \right\}.$$

Set

$$f_{*,k} := \inf\{L(f) \mid L \in \mathcal{X}_k\}.$$

It is not hard to see that the computation of $f_{*,k}$ as a semidefinite programming problem and the sequence $\{f_{*,k}\}_k$ converges monotonically, increasing to the optimal value $f(x^*)$. Moreover, we have the following result.

Theorem 8.11. *Assume that f is coercive on S and that the optimal problem $\inf_{x \in S} f(x)$ has a unique minimizer $x^* \in S$ for which the regularity, strict complementarity and second-order sufficiency conditions. Then there exists an integer number k_* such that for all $k \geq k_*$,*

$$f_{*,k} = \inf_{x \in S} f(x) \quad and \quad (L_{*,k}(\pi_1), \ldots, L_{*,k}(\pi_n)) = x^*,$$

where $\pi_i(x) := x_i$ for $i = 1, \ldots, n$.

Proof. The proof is similar to that of Theorem 7.13 and we leave the details to the reader. □

Bibliographic notes

Most results of Section 8.1 are taken from a paper by Nie, Demmel and Sturmfels [110]. The materials in Sections 8.2–8.4 are taken from papers by Hà and Phạm [51–53]. Theorem 8.11 is derived from results in the papers by Đoạt, Hà and Phạm [32], and by Lee and Phạm [84, 85].

Theorem 8.1 is due to Schweighofer [139]. The proof of Theorem 8.2 presented here follows that in [139, Corollary 47]. To handle the case where no global minimizer exists, Schweighofer [139] uses a sums of squares certificate and the concept of gradient tentacles. There are several recent references on minimizing polynomials by way of the gradients, for example, works by Hanzon and Jibetean [56], Jibetean and Laurent [64], Laurent [81], and Parrilo [120].

Demmel, Nie and Powers [24] generalized the gradient sum of squares relaxation in [110] to find the minimum of a polynomial on a non-compact basic closed semi-algebraic set by using the KKT conditions, under the assumption that one of the global minimizers satisfies the KKT conditions. See also [111] for refinements.

Chapter 9

Convex Polynomial Optimization

Abstract

We show that the Lasserre hierarchy of semidefinite programming relaxations with a slightly extended quadratic module for convex polynomial optimization problems always converges asymptotically even in the case of non-compact semi-algebraic feasible sets. We then prove that the positive definiteness of the Hessian of the associated Lagrangian at a saddle point guarantees the finite convergence of the hierarchy. We show these results by establishing sums of squares certificates for positive (non-negative) convex polynomials over convex semi-algebraic sets.

9.1 Further properties on convex polynomials

In this section, we present some properties of convex polynomials. We begin with the following observation.

Lemma 9.1. *Let $f \in \mathbb{R}[\underline{x}]$ be a convex polynomial which is bounded below on $\mathbb{R}^n$. Then there exist an orthogonal $n \times n$ matrix A and a coercive convex polynomial $g\colon \mathbb{R}^l \to \mathbb{R}$, $1 \leq l \leq n$, such that for all $x = (x_1, \ldots, x_l, \ldots, x_n) \in \mathbb{R}^n$,*

$$f(Ax) = g(x_1, \ldots, x_l).$$

In particular, f attains its infimum on $\mathbb{R}^n$.

Proof. Let

$$E_f := \{d \in \mathbb{R}^n \mid f(x + td) = f(x), \, \forall \, t \in \mathbb{R} \text{ and } \forall \, x \in \mathbb{R}^n\}.$$

It is easy to verify directly that E_f is a subspace of $\mathbb{R}^n$. Let $l :=$ $n - \dim E_f$, and let $e_1, \ldots, e_n \in \mathbb{R}^n$ be an orthonormal basis such that $\operatorname{span}\{e_{l+1}, \ldots, e_n\} = E_f$ and $\operatorname{span}\{e_1, \ldots, e_l\} = E_f^\perp$, where $E_f^\perp$ is the orthogonal complement of E_f. Let $A := [e_1, \ldots, e_n]$. Then A is an orthogonal matrix. Define $g \colon \mathbb{R}^l \to \mathbb{R}$ by $g(x_1, \ldots, x_l) :=$ $f(\sum_{i=1}^l x_i e_i)$. Then g is a convex polynomial which is bounded below on $\mathbb{R}^l$. Moreover, for all $x \in \mathbb{R}^n$, we have

$$f(Ax) = f\left(\sum_{i=1}^n x_i e_i\right) = f\left(\sum_{i=1}^l x_i e_i + \sum_{i=l+1}^n x_i e_i\right)$$

$$= f\left(\sum_{i=1}^l x_i e_i\right) = g(x_1, \ldots, x_l),$$

where the third equality follows by the fact that $\sum_{i=l+1}^n x_i e_i \in E_f$.

To verify that g is indeed coercive, we assume, on the contrary, that there exist a number $\alpha \in \mathbb{R}$ and a sequence $\{a^k\}_k \subset \mathbb{R}^l$ such that $g(a^k) \le \alpha$ for all k and $\|a^k\| \to +\infty$ as $k \to \infty$. Let $a \in \mathbb{R}^l$. Then, by passing to subsequence if necessary, we may assume that $\frac{a^k - a}{\|a^k - a\|} \to v \ne 0$. Let $t \ge 0$. For sufficiently large k, we have $0 <$ $\frac{t}{\|a^k - a\|} < 1$, and so

$$g\left(a + t\frac{a^k - a}{\|a^k - a\|}\right) = g\left(\left(1 - \frac{t}{\|a^k - a\|}\right)a + \frac{t}{\|a^k - a\|}a^k\right)$$

$$\le \left(1 - \frac{t}{\|a^k - a\|}\right)g(a) + \frac{t}{\|a^k - a\|}g(a^k)$$

$$\le \max\{g(a), \alpha\}.$$

Letting $k \to \infty$, we get that $g(a + tv) \le \max\{g(a), \alpha\}$ for all $t \ge 0$. By assumption, g is bounded below. So, $t \mapsto g(a + tv)$ is either a constant or a polynomial with even degree ≥ 2. It follows that g takes a constant value on $\{a + tv \mid t \ge 0\}$ for all $a \in \mathbb{R}^l$. Then, for all $t \ge 0$ and for any $a \in \mathbb{R}^l$, $g(a - tv) = g(a - tv + tv) = g(a)$. Thus,

$$g(a) = g(a + tv) \quad \text{for all } a \in \mathbb{R}^l \text{ and } t \in \mathbb{R}.$$

Let $\tilde{v} := (v, 0, \ldots, 0) \in \mathbb{R}^n$ and $d := A\tilde{v} = \sum_{i=1}^{l} v_i e_i \in E_f^\perp$. Since $v \neq 0$, $d \neq 0$. Moreover, for all $x \in \mathbb{R}^n$ and $t \in \mathbb{R}$,

$$f(x + td) = f(A(A^{-1}x + t\tilde{v})) = g(z + tv) = g(z) = f(x),$$

where $z = ((A^{-1}x)_1, \ldots, (A^{-1}x)_l) \in \mathbb{R}^l$. So, by definition, $d \in E_f$. Consequently, we obtain that $d \in (E_f \cap E_f^\perp) \backslash \{0\}$, which is impossible. Hence, g is coercive.

Since the polynomial g is coercive, there exists $z^* := (z_1^*, \ldots, z_l^*) \in \mathbb{R}^l$ such that $g(z^*) = \inf_{z \in \mathbb{R}^l} g(z)$. Let $x^* := A\binom{z^*}{0} = z_1^* e_1 + \cdots + z_l^* e_l \in E_f^\perp \subset \mathbb{R}^n$. Then, $f(x^*) = \inf_{x \in \mathbb{R}^n} f(x) = g(z^*)$. □

Recall that a real-valued function $f \colon \mathbb{R}^n \to \mathbb{R}$ is called *strictly convex* if for all $x, y \in \mathbb{R}^n, x \neq y$, and all $t \in (0, 1)$,

$$f(tx + (1 - t)y) < tf(x) + (1 - t)f(y).$$

The following lemma on strict convexity and coercivity of convex polynomials plays a key role in proving sums of squares representation results for non-negative convex polynomials and also the finite convergence of Lasserre's hierarchy.

Lemma 9.2. *Let $f \colon \mathbb{R}^n \to \mathbb{R}$ be a convex polynomial. If the Hessian $\nabla^2 f(x^0)$ of f is positive definite at some point $x^0 \in \mathbb{R}^n$, then f is coercive and strictly convex on $\mathbb{R}^n$.*

Proof. (Coercivity) We first prove that f is coercive on $\mathbb{R}^n$. Let M be a real number such that $M > f(x^0)$. To prove coercivity of f on $\mathbb{R}^n$, it suffices to show that the set

$$S := \{x \in \mathbb{R}^n \mid f(x) \leq M\}$$

is compact. On the contrary, suppose that there exists a sequence $\{a^k\}_{k \in \mathbb{N}} \subset S$ such that $\|a^k\| \to \infty$ as $k \to \infty$. Without loss of generality, we may assume that there exists a non-zero vector v such that

$$v := \lim_{k \to \infty} \frac{a^k - x^0}{\|a^k - x^0\|}.$$

Let $t \geq 0$. For sufficiently large k, we have $0 < \frac{t}{\|a^k - x^0\|} < 1$, and so

$$f\left(x^0 + t\frac{a^k - x^0}{\|a^k - x^0\|}\right) = f\left(\left(1 - \frac{t}{\|a^k - x^0\|}\right)x^0 + \frac{t}{\|a^k - x^0\|}a^k\right)$$

$$\leq \left(1 - \frac{t}{\|a^k - x^0\|}\right)f(x^0) + \frac{t}{\|a^k - x^0\|}f(a^k)$$

$$\leq M.$$

Letting $k \to \infty$, we get

$$f(x^0 + tv) \leq M \quad \text{for all } t \geq 0.$$

On the other hand, we have for all $t \in \mathbb{R}$,

$$f(x^0 + tv) = f(x^0) + \langle \nabla f(x^0), v \rangle t + \tfrac{1}{2}\langle \nabla^2 f(x^0)v, v \rangle t^2$$

$$+ \text{ higher order terms in } t.$$

Since the Hessian $\nabla^2 f(x^0)$ is positive definite, $\langle \nabla^2 f(x^0)v, v \rangle > 0$. Therefore, the one-dimensional convex polynomial function

$$\mathbb{R} \to \mathbb{R}, \quad t \mapsto f(x^0 + tv),$$

is of degree ≥ 2, and so its degree is even ≥ 2. This is a contradiction since $f(x^0 + tv) \leq M$ for all $t \geq 0$.

(Strict convexity) Suppose on the contrary that f is not strictly convex. Then, there exist $x, y \in \mathbb{R}^n, x \neq y$, and $t_0 \in (0, 1)$ such that

$$f((1 - t_0)x + t_0 y) = (1 - t_0)f(x) + t_0 f(y).$$

Define the function $\varphi \colon [0, 1] \to \mathbb{R}$ by $\varphi(t) := f((1 - t)x + ty) - (1 - t)f(x) - tf(y)$. Then, φ is a convex polynomial, $\varphi(t) \leq 0$ for each $t \in [0, 1]$, and $\varphi(t_0) = 0 = \max_{t \in [0,1]} \varphi(t)$. As φ is a convex function on $[0, 1]$, it attains its maximum on the extreme points of $[0, 1]$. Therefore $\varphi(t) = 0$ for all $t \in [0, 1]$, or equivalently,

$$f((1 - t)x + ty) = (1 - t)f(x) + tf(y), \quad \forall t \in [0, 1].$$

Now, define a polynomial ψ on $\mathbb{R}$ by $\psi(\lambda) := f(x + \lambda(y - x))$, $\lambda \in \mathbb{R}$. Clearly, ψ is affine on $[0, 1]$, and moreover, it is coercive on $\mathbb{R}$

because f is coercive on $\mathbb{R}^n$. We show that ψ is indeed affine over $\mathbb{R}$. To do this, let the degree of the one-dimensional polynomial ψ be d. Then, for each $\lambda \in \mathbb{R}$,

$$\psi(\lambda) = \psi(0) + \psi'(0)\lambda + \frac{\psi''(0)}{2}\lambda^2 + \cdots + \frac{\psi^{(d)}(0)}{d!}\lambda^d.$$

As ψ is affine over $[0,1]$, $\psi^{(i)}(0) = 0$ for $i = 2, \ldots, d$, and so, $\psi(\lambda) = \psi(0) + \psi'(0)\lambda$. Hence, ψ is affine over $\mathbb{R}$. This contradicts the fact that ψ is coercive on $\mathbb{R}$. $\qquad\square$

9.2 Asymptotic convergence

Let $f, h_1, \ldots, h_m \in \mathbb{R}[\underline{x}]$ be convex polynomials and assume that

$$S := \{x \in \mathbb{R}^n \mid h_1(x) \leq 0, \ldots, h_m(x) \leq 0\} \neq \emptyset.$$

In this section, we will construct a sequence of semidefinite programming relaxations to the problem

$$f_* := \inf_{x \in S} f(x).$$

To do this we will establish a sums of squares certificate for positivity of convex polynomials over (not necessarily compact) convex semi-algebraic sets. We begin with the following observation.

Lemma 9.3. *Let*

$$\Omega := \{(y_1, \ldots, y_{m+1}) \in \mathbb{R}^{m+1} \mid \text{there exists } x \in \mathbb{R}^n \text{ such that}$$
$$h_j(x) \leq y_j, j = 1, \ldots, m, \ f(x) \leq y_{m+1}\}.$$

Then Ω is a non-empty closed convex set.

Proof. It is clear that Ω is a non-empty convex set. To see the closeness of Ω, take any sequence $\{(y_1^k, \ldots, y_{m+1}^k)\}_{k \in \mathbb{N}} \subset \Omega$ such that $(y_1^k, \ldots, y_{m+1}^k) \to (y_1, \ldots, y_{m+1})$ as $k \to \infty$. By the definition of Ω, for each k, there exists $x^k \in \mathbb{R}^n$ such that $f(x) \leq y_{m+1}^k$ and

$h_j(x^k) \le y_j^k, j = 1, \ldots, m$. Consider the optimization problem

$$P_* := \inf_{x, z_1, \ldots, z_{m+1}} \sum_{j=1}^{m+1} (z_j - y_j)^2$$

subject to $h_j(x) - z_j \le 0, j = 1, \ldots, m, \ f(x) - z_{m+1} \le 0.$

This is a convex polynomial program with the optimal value $P_* = 0$ because it holds that

$$0 \le P_* \le \sum_{j=1}^{m+1} (y_j^k - y_j)^2 \to 0 \quad \text{as } k \to \infty.$$

Moreover, Theorem 4.1 implies that the infimum of the problem is attained. Hence, there exists $x \in \mathbb{R}^n$ such that $h_j(x) \le y_j$ and $f(x) \le y_{m+1}$. So $(y_1, \ldots, y_{m+1}) \in \Omega$. Therefore, Ω is closed. $\square$

The following result is a dual characterization for positivity of f on S.

Lemma 9.4 (Convex Farkas Lemma I). *The following statements are equivalent:*

(i) $f(x) > 0$ *for all* $x \in S$.
(ii) *There exists* $\lambda \in \mathbb{R}_+^m$ *such that* $f(x) + \sum_{j=1}^m \lambda_j h_j(x) > 0$ *for all* $x \in \mathbb{R}^n$.

Proof.
   (ii) $\Rightarrow$ (i) This is straightforward and is left as an exercise.
   (i) $\Rightarrow$ (ii) Thanks to Theorem 4.1, there exists $x^* \in S$ such that $f_* := \inf_{x \in S} f(x) = f(x^*)$. The assumption gives $f_* > 0$. Let

$$\Omega := \{(y_1, \ldots, y_{m+1}) \in \mathbb{R}^{m+1} \mid \text{ there exists } x \in \mathbb{R}^n \text{ such that}$$

$$h_j(x) \le y_j, j = 1, \ldots, m, \ f(x) \le y_{m+1}\}.$$

By Lemma 9.3, Ω is a non-empty closed convex set. Clearly, $(0, \ldots, 0, \frac{f_*}{2}) \notin \Omega$. So, from the strict separation theorem, there exist a number $\alpha \in \mathbb{R}$ and a non-zero vector $(\lambda, \lambda_{m+1}) \in \mathbb{R}^m \times \mathbb{R}$ such that

$$\lambda_{m+1} \frac{f_*}{2} \le \alpha < \sum_{j=1}^{m+1} \lambda_j y_j \quad \text{for all } y := (y_1, \ldots, y_{m+1}) \in \Omega.$$

Let $e_1, \ldots, e_{m+1}$ be the canonical basic vectors in $\mathbb{R}^{m+1}$ and fix $y \in \Omega$. Then for all $k = 1, \ldots, m+1$, and all $R > 0$, $y + Re_k \in \Omega$. Hence

$$\alpha < \sum_{j=1}^{m+1} \lambda_j y_j + \lambda_k R.$$

Letting $R \to +\infty$ yields $\lambda_k \geq 0$ for all $k = 1, \ldots, m+1$.

Suppose now that $\lambda_{m+1} = 0$. Note that $(h_1(x), \ldots, h_m(x), f(x)) \in \Omega$ for all $x \in \mathbb{R}^n$. Hence,

$$0 \leq \alpha < \sum_{j=1}^{m} \lambda_j h_j(x) \quad \text{for all } x \in \mathbb{R}^n.$$

This contradicts the consistency assumption that $S \neq \emptyset$. So, without loss of generality, we can assume that $\lambda_0 = 1$ and we get that

$$f(x) + \sum_{j=1}^{m} \lambda_j h_j(x) > \frac{f_*}{2} > 0 \quad \text{for all } x \in \mathbb{R}^n. \qquad \square$$

In what follows, we fix a real number M such that $M > f(x^0)$ for some $x^0 \in S$. Recall that the quadratic module $\mathbf{Q}(-h_1, \ldots, -h_m, M-f)$ generated by the polynomials $-h_1, \ldots, -h_m$ and $M-f$ is, by definition, the set

$$\left\{ \sigma_0 - \sum_{j=1}^{m} \sigma_j h_j + \sigma(M - f) \, | \, \sigma, \sigma_0, \sigma_1, \ldots, \sigma_m \in \sum \mathbb{R}[x]^2 \right\}.$$

The following result is a sums of squares certificate for positivity of f on S which is suitable for optimization purposes.

Theorem 9.1. *If $f > 0$ on S, then $f \in \mathbf{Q}(-h_1, \ldots, -h_m, M-f)$.*

Proof. Thanks to Lemma 9.4, there exists $\lambda \in \mathbb{R}_+^m$ such that

$$f(x) + \sum_{j=1}^{m} \lambda_j h_j(x) > 0 \quad \text{for all } x \in \mathbb{R}^n.$$

Define the polynomial $\widetilde{f} \colon \mathbb{R}^n \to \mathbb{R}, x \mapsto \widetilde{f}(x)$, by

$$\widetilde{f}(x) := f(x) + \sum_{j=1}^{m} \lambda_j h_j(x).$$

Clearly, $\widetilde{f}$ is convex and positive on $\mathbb{R}^n$. Lemma 9.1 shows that there exist an orthogonal $n \times n$ matrix A and a coercive polynomial $\widetilde{g} \colon \mathbb{R}^l \to \mathbb{R}$ such that for all $x = (x_1, \ldots, x_l, \ldots, x_n) \in \mathbb{R}^n$,

$$\widetilde{f}(Ax) = \widetilde{g}(x_1, \ldots, x_l). \tag{9.1}$$

Note that the set $\{x \in \mathbb{R}^n \mid M - \widetilde{f}(x) \geq 0\}$ is non-empty because it contains the point $x^0 \in S$. As $\widetilde{g}$ is coercive on $\mathbb{R}^l$,

$$T := \{x_1, \ldots, x_l) \in \mathbb{R}^l \mid M - \widetilde{g}(x_1, \ldots, x_l) \geq 0\}$$

is a non-empty compact set. Since $\widetilde{f} > 0$ on $\mathbb{R}^n$, it follows from (9.1) that $\widetilde{g} > 0$ on $\mathbb{R}^l$, and in particular $\widetilde{g} > 0$ on T.

Now, from Theorem 7.3 (Schmüdgen's Positivstellensätz), we get that there exist sums of squares σ_0, σ_1 over $\mathbb{R}^l$ such that

$$\widetilde{g} = \sigma_0 + \sigma_1(M - \widetilde{g}).$$

From (9.1), we see that, for each $x = (x_1, \ldots, x_l, x_{l+1}, \ldots, x_n) \in \mathbb{R}^n$, $\widetilde{f}(Ax) = \widetilde{g}(x_1, \ldots, x_l)$. So, for each $x = (x_1, \ldots, x_l, x_{l+1}, \ldots, x_n) \in \mathbb{R}^n$,

$$\widetilde{f}(Ax) = \sigma_0(x_1, \ldots, x_l) + \sigma_1(x_1, \ldots, x_l)(M - \widetilde{f}(Ax)).$$

Then, for each $z \in \mathbb{R}^n$,

$$\widetilde{f}(z) = \sigma_0\big((A^{-1}z)_1, \ldots, (A^{-1}z)_l\big)$$
$$+ \sigma_1\big((A^{-1}z)_1, \ldots, (A^{-1}z)_l\big)(M - \widetilde{f}(z)).$$

Using the definition of $\widetilde{f}$, we obtain that, for each $z \in \mathbb{R}^n$,

$$f(z) + \sum_{j=1}^{m} \lambda_j h_j(z)$$
$$= \sigma_0\big((A^{-1}z)_1, \ldots, (A^{-1}z)_l\big)$$
$$+ \sigma_1\big((A^{-1}z)_1, \ldots, (A^{-1}z)_l\big)\left(M - f(z) - \sum_{j=1}^{m} \lambda_j h_j(z)\right).$$

Thus, for each $z \in \mathbb{R}^n$,

$$f(z) = \sigma_0\big((A^{-1}z)_1, \ldots, (A^{-1}z)_l\big)$$
$$+ \sigma_1\big((A^{-1}z)_1, \ldots, (A^{-1}z)_l\big)(M - f(z))$$
$$- \sum_{j=1}^{m} \big(\sigma_1\big((A^{-1}z)_1, \ldots, (A^{-1}z)_l\big)\lambda_j + \lambda_j\big)h_j(z),$$

where $z \mapsto \sigma_i\big((A^{-1}z)_1, \ldots, (A^{-1}z)_l\big)$, $i = 0, 1$, are sums of squares and $\lambda_j \geq 0$, for $j = 1, \ldots, m$. □

Assume that f is bounded from below on S and consider the problem:

$$f_* := \inf\{f(x) \mid x \in S\}.$$

We now construct a sequence of semidefinite programs whose optimal values converge monotonically, increasing to the optimal value f_*. To do this, for each integer $k \geq \frac{1}{2}\max\{\deg f, \deg h_1, \ldots, \deg h_m\}$, we will denote by $\mathbf{Q}_k(-h_1, \ldots, -h_m, M - f)$ the set

$$\left\{ \sigma_0 - \sum_{j=1}^{m} \sigma_j h_j + \sigma(M - f) \,\middle|\, \sigma, \sigma_0, \sigma_1, \ldots, \sigma_m \in \sum \mathbb{R}[x]^2, \right.$$

$$\left. \deg \sigma_0 \leq 2k, \deg \sigma_j h_j \leq 2k, \deg \sigma(M - f) \leq 2k \right\}$$

which is the $2k$th truncated quadratic module generated by $-h_1, \ldots, -h_m, M - f$. A lower bound for f_* can be found by solving the sequence of semidefinite programming relaxations:

$$\overline{f}_{k,*} := \sup_{r \in \mathbb{R}}\{r \mid f - r \in \mathbf{Q}_k(-h_1, \ldots, -h_m, M - f)\}.$$

As shown in Proposition 7.1, the problem of computing the supremum $\overline{f}_{*,k}$ can be reduced to a semidefinite program. Furthermore, it can be easily verified that

$$\overline{f}_{*,k} \leq \overline{f}_{*,k+1} \leq \cdots \leq f_*.$$

Theorem 9.2. *Assume that f is bounded from below on S. Then,* $\lim_{k \to \infty} \overline{f}_{*,k} = f_*.$

Proof. Fix $\varepsilon > 0$ arbitrary. We have $f - f_* + \varepsilon > 0$ on S. Theorem 9.1 shows that $f - f_* + \varepsilon \in \mathbf{Q}(-h_1, \ldots, -h_m, M - f)$. So there exists an integer $k(\varepsilon)$ such that

$$\overline{f}_{*,k} \geq f_* - \varepsilon \quad \text{for all } k \geq k(\varepsilon).$$

Therefore, as $\varepsilon > 0$ is arbitrary, letting $\varepsilon \to 0^+$ yields the desired result. $\qquad\square$

9.3   Finite convergence

Let $f, h_1, \ldots, h_m \in \mathbb{R}[x]$ be convex polynomials and assume that

$$S := \{x \in \mathbb{R}^n \mid h_1(x) \leq 0, \ldots, h_m(x) \leq 0\} \neq \emptyset.$$

Definition 9.1. We say that the *Slater condition* holds if and only if there exists a point $x^0 \in \mathbb{R}^n$ such that $h_j(x^0) < 0$ for all $j = 1, \ldots, m$.

The following lemma gives a dual characterization of non-negativity of f on S under the Slater condition.

Lemma 9.5 (Convex Farkas Lemma II). *Suppose that the Slater condition holds. Then the following statements are equivalent:*

(i) $f(x) \geq 0$ *for all* $x \in S$.

(ii) *There exists* $\lambda \in \mathbb{R}_+^m$ *such that* $f(x) + \sum_{j=1}^m \lambda_j h_j(x) \geq 0$ *for all* $x \in \mathbb{R}^n$.

Proof. (ii) $\Rightarrow$ (i) This implication trivially holds by the construction.

(i) $\Rightarrow$ (ii) Suppose that f is non-negative on S. Then, by Lemma 9.4, for each $\epsilon > 0$ there exists $(\lambda_1^\epsilon, \ldots, \lambda_m^\epsilon) \in \mathbb{R}_+^m$ such that

$$f(x) + \sum_{j=1}^m \lambda_j^\epsilon h_j(x) + \epsilon > 0, \quad \forall x \in \mathbb{R}^n.$$

We first observe that $\{\lambda_j^\epsilon\}$ is bounded. Otherwise, we assume that $\lim_{\epsilon \to 0^+} \lambda_i^\epsilon = +\infty$ for some i. Then

$$\lim_{\epsilon \to 0^+} \left[f(x^0) + \sum_{j=1}^m \lambda_j^\epsilon h_j(x^0) + \epsilon \right] = -\infty,$$

where the limit follows as $h_j(x^0) < 0$ for all $j = 1, \ldots, m$. This is impossible and so, λ_j^ϵ are all bounded. By passing to subsequences if necessary, we may assume that $\lambda_j^\epsilon \to \lambda_j$ for some $\lambda_j \geq 0$. Then it is easy to obtain (ii). □

Consider the convex polynomial optimization problem

$$f_* := \inf_{x \in S} f(x). \tag{9.2}$$

In this section, under certain conditions, we show that there is finite convergence of the bounds $\overline{f}_{*,k}$ constructed in the previous section to the infimum value f_*. To do this we will present representation results for non-negativity of convex polynomials over convex semi-algebraic sets.

Definition 9.2. The *Lagrangian function* of problem (9.2) is defined as

$$\mathscr{L}(x, \lambda) := f(x) + \sum_{j=1}^m \lambda_j h_j(x), \quad (x, \lambda) \in \mathbb{R}^n \times \mathbb{R}_+^m.$$

A pair $(x^*, \lambda^*) \in \mathbb{R}^n \times \mathbb{R}_+^m$ is called a *saddle point* of the Lagrangian function $\mathscr{L}$ if

$$\forall (x, \lambda) \in \mathbb{R}^n \times \mathbb{R}_+^m, \quad \mathscr{L}(x, \lambda^*) \geq \mathscr{L}(x^*, \lambda^*) \geq \mathscr{L}(x^*, \lambda). \tag{9.3}$$

Remark 9.1. Assume that f is bounded from below on S. It follows from Theorem 4.1 that f attains its infimum on S at some point $x^* \in S$. Assume moreover, that the Slater condition holds. Then, by Lemma 9.5, there exists $\lambda^* \in \mathbb{R}_+^m$ such that (x^*, λ^*) is a saddle point of the Lagrangian function $\mathscr{L}$.

Theorem 9.3. *If the Lagrangian function $\mathscr{L}$ has a saddle point $(x^*, \lambda^*) \in S \times \mathbb{R}_+^m$ with $\nabla_x^2 \mathscr{L}(x^*, \lambda^*) \succ 0$, then, for any $M \in \mathbb{R}$ with $M > f(x^*)$, we have*

$$f - f(x^*) \in \mathbf{Q}(-h_1, \ldots, -h_m, M - f),$$

where $\nabla_x^2 \mathscr{L}(x^*, \lambda^*)$ stands for the Hessian of $\mathscr{L}$ with respect to x.

Proof. Since (x^*, λ^*) is a saddle point of the Lagrangian function $\mathscr{L}$ and $x^* \in S$, it follows that $\lambda_j^* h_j(x^*) = 0$ for $j = 1, \ldots, m$, and that x^* is a minimizer of f over S. Define the polynomial $\widetilde{f} \colon \mathbb{R}^n \to \mathbb{R}, x \mapsto \widetilde{f}(x)$, by

$$\widetilde{f}(x) := f(x) - f(x^*) + \sum_{j=1}^{m} \lambda_j^* h_j(x).$$

It is easy to check that the polynomial $\widetilde{f}$ is convex and non-negative on $\mathbb{R}^n$. Moreover, we have $\widetilde{f}(x^*) = 0, \nabla \widetilde{f}(x^*) = 0,$ and $\nabla^2 \widetilde{f}(x^*) \succ 0$. Now, Lemma 9.2 shows us that the polynomial $\widetilde{f}$ is strictly convex and coercive, which implies that x^* is the unique minimizer of $\widetilde{f}$ on $\mathbb{R}^n$ and that

$$\{x \in \mathbb{R}^n \mid -\widetilde{f}(x) + M - f(x^*) \geq 0\}$$

is a compact set with x^* being its interior point. We next apply Theorems 7.7 and 7.8 to conclude that there exist sums of squares σ_0, σ_1 such that, for each $x \in \mathbb{R}^n$,

$$\widetilde{f}(x) = \sigma_0(x) + \sigma_1(x)(-\widetilde{f}(x) + M - f(x^*)).$$

So, for each $x \in \mathbb{R}^n$,

$$f(x) - f(x^*) = \sigma_0 - \sum_{j=1}^{m} (\lambda_j^* + \lambda_j^* \sigma_1) h_j(x) + \sigma_1(M - f(x)).$$

Hence, the conclusion follows. □

Remark 9.2. Assume that the quadratic module $\mathbf{Q}(-h_1, \ldots, -h_m)$ is Archimedean. Then the set S is compact and so f attains its infimum on S at some point $x^* \in S$. Assume moreover, that the Slater condition holds. Then Lemma 9.5 shows that there exists $\lambda^* \in \mathbb{R}_+^m$ such that (x^*, λ^*) is a saddle point of the Lagrangian function $\mathscr{L}$. Take a real number M such that $M - f > 0$ on S. By Putinar's Positivstellensätz (Theorem 7.4), $M - f \in \mathbf{Q}(-h_1, \ldots, -h_m)$. This, together with Theorem 9.3, implies easily that if $\nabla_x^2 \mathscr{L}(x^*, \lambda^*) \succ 0$ then

$$f - f(x^*) \in \mathbf{Q}(-h_1, \ldots, -h_m).$$

Note that the positive definiteness of $\nabla_x^2 \mathscr{L}(x^*, \lambda^*)$ is guaranteed if the Hessian $\nabla^2 f(x^*)$ is positive definite.

The following simple one-dimensional example illustrates that our representation result can be applied to the case where the Hessian $\nabla^2 f$ is not positive definite at a minimizer.

Example 9.1. Let $f(x) = x$ and $h(x) = x^2 - 1$. Then, $S := \{x \in \mathbb{R} \mid h(x) \leq 0\} = [-1, 1]$. Clearly, $x^* := -1$ is the unique minimizer of f on S and $\nabla^2 f(x^*)$ is not positive definite. On the other hand, direct verification shows that $(x^*, \lambda^*) := (-1, \frac{1}{2})$ is a saddle point of the Lagrangian function $\mathscr{L}(x, \lambda) := f(x) + \lambda h(x) = x + \lambda(x^2 - 1)$, and $\nabla_x^2 \mathscr{L}(x^*, \lambda^*) \succ 0$. Moreover, the Slater condition is satisfied and the quadratic module $\mathbf{Q}(-h)$ is Archimedean. So, it follows from the previous remark that $f - f(x^*) = f + 1 \in \mathbf{Q}(-h)$. Indeed, $f(x) - f(x^*) = x + 1 = \frac{1}{2}(x + 1)^2 + \frac{1}{2}(1 - x^2) \in \mathbf{Q}(-h)$.

As we see in the following theorem, under the Slater condition and the positive definiteness of the Hessian of f at a minimizer, we obtain a sharper representation than the one in Theorem 9.3.

Theorem 9.4. *Assume that the Slater condition holds. If f attains its infimum on S at some point $x^* \in S$ with $\nabla^2 f(x^*) \succ 0$ then, for any $M > f(x^*)$, there exist a vector $\lambda \in \mathbb{R}_+^m$ and sums of squares σ_0, σ_1 such that*

$$f - f(x^*) = \sigma_0 - \sum_{j=1}^m \lambda_j^* h_j + \sigma_1(M - f).$$

Proof. The Slater condition and Lemma 9.5 guarantee that there exists a vector $\lambda^* \in \mathbb{R}_+^m$ such that (x^*, λ^*) is a saddle point of the Lagrangian function $\mathscr{L}(x, \lambda) := f(x) + \sum_{j=1}^m \lambda_j h_j(x)$. Define the polynomial $\widetilde{f} \colon \mathbb{R}^n \to \mathbb{R}, x \mapsto \widetilde{f}(x)$, by

$$\widetilde{f}(x) := f(x) - f(x^*) + \sum_{j=1}^m \lambda_j^* h_j(x).$$

It is easy to check that the polynomial $\widetilde{f}$ is convex and non-negative on $\mathbb{R}^n$. Moreover, we have $\widetilde{f}(x^*) = 0, \nabla \widetilde{f}(x^*) = 0$, and $\nabla^2 \widetilde{f}(x^*) \succ 0$.

Now, as $\nabla^2 f(x^*) \succ 0$, Lemma 9.2 shows that the polynomial f is coercive, and hence $T := \{x \in \mathbb{R}^n \mid M - f(x) \geq 0\}$ is a non-empty compact set with x^* being its interior point.

Applying Theorems 7.7 and 7.8 to the problem $\min_{x \in T} \widetilde{f}(x)$, we conclude that there exist sums of squares σ_0, σ_1 such that, for each $x \in \mathbb{R}^n$,

$$\widetilde{f}(x) = \sigma_0(x) + \sigma_1(x)(M - f(x)).$$

So, for each $x \in \mathbb{R}^n$,

$$f(x) - f(x^*) = \sigma_0(x) - \sum_{j=1}^{m} \lambda_j^* h_j(x) + \sigma_1(x)(M - f(x)).$$

Hence, the conclusion follows. $\qquad\qquad\qquad\qquad\qquad\qquad \square$

Let $M \in \mathbb{R}$ be such that $M > f(x^0)$ for some $x^0 \in S$. We have

$$f_* = \inf_{x \in S, M - f(x) \geq 0} f(x).$$

This leads to consider the sequence of semidefinite programming relaxations:

$$\overline{f}_{k,*} := \sup_{r \in \mathbb{R}} \{r \mid f - r \in \mathbf{Q}_k(-h_1, \ldots, -h_m, M - f)\}, \qquad (9.4)$$

where $\mathbf{Q}_k(-h_1, \ldots, -h_m, M - f)$ stands for the $2k$th truncated quadratic module generated by $-h_1, \ldots, -h_m, M - f$. As shown in Theorem 9.2, the sequence $\{\overline{f}_{*,k}\}_{k \in \mathbb{N}}$ converges monotonically increasing to f_*. Moreover, we have the following finite convergence result.

Theorem 9.5. *Assume that the Lagrangian function $\mathscr{L}$ has a saddle point $(x^*, \lambda^*) \in S \times \mathbb{R}_+^m$ with $\nabla_x^2 \mathscr{L}(x^*, \lambda^*) \succ 0$. Then there exists an integer k_* such that $\overline{f}_{*,k} = f_*$ for all $k \geq k_*$.*

Proof. As $(x^*, \lambda^*) \in S \times \mathbb{R}_+^m$ is a saddle point of $\mathscr{L}$, x^* is minimizer of (9.2) and $f_* = f(x^*)$. Theorem 9.3 shows that $f - f_* \in \mathbf{Q}(-h_1, \ldots, -h_m, M - f)$. So there exists an integer k_* such that

$$\overline{f}_{*,k} \geq f_* \quad \text{for all } k \geq k_*.$$

On the other hand, the sequence $\{\overline{f}_{*,k}\}_k$ is monotonically increasing and bounded from above by f_*. Therefore, $\overline{f}_{*,k} = f_*$ for all $k \geq k_*$.

$$\square$$

Remark 9.3. It is worth noting that in the case where $\nabla^2 f(x^*)$ is positive definite at a minimizer x^* of Problem (9.2), using Theorem 9.4, one can establish finite convergence of a sharper form of approximation problem (9.4), where σ_j, $j = 1, \ldots, m$, in the $2k$-truncated quadratic module $\mathbf{Q}_k(-h_1, \ldots, -h_m, M - f)$ are replaced by the Lagrange multipliers, λ_j^*, $j = 1, \ldots, m$, associated with the minimizer x^*.

The next result shows that the existence of a saddle point of the Lagrangian function of (9.2) at each minimizer is necessary for finite convergence of our hierarchy of semidefinite programming relaxations whenever (9.4) achieves its optimal value.

Proposition 9.1. *Suppose that (9.4) achieves its optimal value. If the saddle point condition (9.3) fails at a minimizer of (9.2), then the hierarchy of relaxations (9.4) does not have finite convergence.*

Proof. By contrary, assume that the hierarchy of semidefinite programming relaxations (9.4) has finite convergence. Let $x^* \in S$ with $f_* := f(x^*) = \min_{x \in S} f(x)$. Then, x^* is a minimizer of the following convex program:

$$\min_{x \in \mathbb{R}^n} \{ f(x) \mid h_j(x) \leq 0, \ j = 1, \ldots, m, \ f(x) - M \leq 0 \}.$$

It follows from Proposition 7.2 that the KKT optimality conditions hold for this program at x^*. Thus, there exist $\lambda^* \in \mathbb{R}_+^m$ and $\gamma^* \in \mathbb{R}_+$ such that

$$\nabla f(x^*) + \sum_{j=1}^{m} \lambda_j^* \nabla h_j(x^*) + \gamma^* \nabla f(x^*) = 0,$$

$$\lambda_j^* h_j(x^*) = 0, \ j = 1, \ldots, m, \quad \text{and} \quad \gamma^*(f(x^*) - M) = 0.$$

As $M > f(x^0) \geq f(x^*)$, we obtain $\gamma^* = 0$. So,

$$\nabla f(x^*) + \sum_{j=1}^{m} \lambda_j^* \nabla h_j(x^*) = 0.$$

Hence, by convexity of $f + \sum_{j=1}^{m} \lambda_j^* h_j$, we get that, for each $x \in \mathbb{R}^n$,

$$f(x) + \sum_{j=1}^{m} \lambda_j^* h_j(x) \geq f(x^*) + \sum_{j=1}^{m} \lambda_j^* h_j(x^*).$$

It is now easy to check that (x^*, λ^*) is a saddle point of the Lagrangian function of (9.2), which is a contradiction.   □

Bibliographic notes

The material in this chapter is taken from a paper by Jeyakumar, Li and Phạm [63].

It is shown by Lasserre in [78] that the Lasserre hierarchy of semidefinite programming approximations to convex polynomial optimization problems is known to converge finitely under some assumptions. Klerk and Laurent [69] give a new proof of the finite convergence property, that does not require the assumption that the Hessian of the objective being positive definite on the entire feasible set, but only at the optimal solution.

The reader is invited to consult the monograph [12] for fundamental aspects of convex algebraic geometry, which is an evolving subject area arising from a synthesis of ideas and techniques from optimization, convex geometry, and algebraic geometry.

Lemma 9.1 can also be proved by the methods in [147]. The results concerning the Slater condition, the convex Farkas lemma, and the existence of saddle points of the Lagrangian function can be found in the book [11].

Glossary

Standard Notation

We denote by $\mathbb{Z}, \mathbb{N}, \mathbb{R}$ and $\mathbb{R}_+$ the sets of integers, non-negative integers, real numbers and non-negative real numbers, respectively.

For $x \in \mathbb{R}^n$, x^T denotes the *transpose* of x. While doing matrix computations, we will view x as a column vector. We denote by $\mathbb{R}[x_1, \ldots, x_n]$ (or $\mathbb{R}[x]$, if $x = (x_1, \ldots, x_n)$) the ring of polynomials in the variables $x_1, \ldots, x_n$ with coefficients in $\mathbb{R}$. If f is a polynomial in $\mathbb{R}[x]$, $\deg f$ denotes its degree.

- $x^\alpha := x_1^{\alpha_1} \cdots x_n^{\alpha_n}$ and $|\alpha| := \sum_{i=1}^n \alpha_i$ for $x = (x_1, \ldots, x_n) \in \mathbb{R}^n$ and $\alpha = (\alpha_1, \ldots, \alpha_n) \in \mathbb{N}^n$
- $\langle x, y \rangle := \sum_{i=1}^n x_i y_i$, the scalar product of $x = (x_1, \ldots, x_n), y = (y_1, \ldots, y_n) \in \mathbb{R}^n$
- $\|x\| := \sqrt{\langle x, x \rangle}$, the Euclidean norm of x
- $\mathbb{S}^{n-1}$, standard sphere (with center 0 and radius 1) in $\mathbb{R}^n$
- $\overline{\mathbb{B}}$, closed unit ball with center 0 in $\mathbb{R}^n$
- $\overline{S}$, closure of a set $S \subset \mathbb{R}^n$
- S^c, complement of a set $S \subset \mathbb{R}^n$
- We adhere to the convention that the supremum (respectively, infimum) of an empty set is equal to $-\infty$ (respectively, $+\infty$).

Special Notation

The following entries are listed in order of appearance.

Chapter 1:

- $\text{dist}(x, S)$, distance from x to S
- graph f, graph of f
- $\pi \colon \mathbb{R}^n \times \mathbb{R}^m \to \mathbb{R}^n$, projection on the first n coordinates
- $\dim S$, dimension of a semi-algebraic set S
- $\nabla f(x)$, gradient of f at x
- $K_0(f)$, set of critical values of f
- $\mathscr{R}(n, d)$, D'Acunto–Kurdyka function

Chapter 2:

- $J(x)$, set of active constraint indices at x
- $\Sigma(f)$, set of critical points of f
- $\Sigma(f, S)$, set of critical points of f on S
- $K_0(f)$, set of critical values of f
- $K_0(f, S)$, set of critical values of f on S
- $\Gamma(f)$, tangency variety of f
- $\Gamma(f, S)$, tangency variety of f on S
- $T_\infty(f)$, set of tangency values of f
- $T_\infty(f, S)$, set of tangency values of f on S

Chapter 3:

- $\hat{\partial} f(x)$, Fréchet subdifferential of f at x
- $\partial f(x)$, limiting subdifferential of f at x
- $\mathfrak{m}_f(x)$, non-smooth slope of f at x
- $[r]_+ := \max\{r, 0\}$
- $0 < r \ll 1$, r is positive and sufficiently small
- $r \gg 1$, r is sufficiently large
- f_Δ, principal part of f at infinity with respect to Δ
- $\mathcal{N}(f)$, Newton polyhedron at infinity of f
- $\mathcal{N}_\infty(f)$, Newton boundary at infinity of f

Chapter 4:

- $0^+ Z$, recession cone of a convex set Z
- ν_F, Rabier function
- $\widetilde{K}_\infty(F)$, set of asymptotic critical values of F

Chapter 5:

- $\mathcal{A}_\mathcal{N}$, subset of polynomials f with $\mathcal{N}(f) \subseteq \mathcal{N}$
- $vec(x)$, vector of monomials corresponding x
- $P_\mathcal{N}$, semi-algebraic function corresponding to a Newton polyhedron $\mathcal{N}$

Chapter 6:

- $\#X$, cardinality of a finite set X
- $KKT(u)$, Karush, Kuhn and Tucker set-valued map
- $\nabla^2 f(x)$, Hessian of f at x
- $T_x S$, generalized tangent space of S at x

Chapter 7:

- $\text{Mat}(s)$, set of all $s \times s$ matrices
- $\text{Sym}(s)$, set of symmetric $s \times s$ matrices
- $\text{Tr}(A)$, trace of a square matrix A
- $A \succeq 0$ ($A \succ 0$), A is positive semidefinite (definite) matrix
- SOS, sum of squares
- $\sum \mathbb{R}[x]^2$, subset of $\mathbb{R}[x]$ consisting of sums of squares
- $\mathbf{I}(g_1, \ldots, g_l)$, ideal generated by polynomials $g_1, \ldots, g_l$ in $\mathbb{R}[x]$
- $\mathbf{P}(h_1, \ldots, h_m)$, preordering generated by polynomials $h_1, \ldots, h_m$ in $\mathbb{R}[x]$
- $\mathbf{Q}(h_1, \ldots, h_m)$, quadratic module generated by polynomials $h_1, \ldots, h_m$ in $\mathbb{R}[x]$
- $\mathbf{I}_k(g_1, \ldots, g_l) \subset \mathbb{R}[x]$, truncated ideal
- $\mathbf{P}_k(h_1, \ldots, h_m) \subset \mathbb{R}[x]$, truncated preordering
- $\mathbf{Q}_k(h_1, \ldots, h_m) \subset \mathbb{R}[x]$, truncated quadratic module
- $\mathcal{Z}(g_1, \ldots, g_l)$, set of zeros of polynomials $g_1, \ldots, g_l$

Chapter 8:

- $R_\infty(f, S)$, set of asymptotic values of f on S
- $\Gamma_M(f)$, truncated tangency variety of f
- $\Gamma_M(f, S)$, truncated tangency variety of f on S
- $\mathbf{I}(f, S)$, ideal corresponding to $\Sigma(f, S)$ or $\Gamma(f, S)$
- $\mathbf{I}_M(f, S)$, truncated ideal

Chapter 9:

- $\mathscr{L}(x, \lambda)$, Lagrange function
- $\mathbf{Q}(h_1, \ldots, h_m, M - f)$, quadratic module generated by polynomials $h_1, \ldots, h_m$ and $M - f$
- $\mathbf{Q}_k(h_1, \ldots, h_m, M - f) \subset \mathbb{R}[x]$, truncated quadratic module

Bibliography

1. D'Acunto D, Kurdyka K. Explicit bounds for the Lojasiewicz exponent in the gradient inequality for polynomials. *Ann. Pol. Math.* 2005; 87: 51–61.

2. Alizadeh F, Haeberly JPA, Overton ML. Complementarity and nondegeneracy in semidefinite programming. *Math. Program. Ser. B* 1977; 77(2): 111–128.

3. Andronov VG, Belousov EG, Shironin VM. On solvability of the problem of polynomial programming. *Izvestija Akadem. Nauk SSSR, Tekhnicheskaja Kibernetika* 1982; 4: 194–197

4. Anjos MF, Lasserre JB. (eds) *Handbook on semidefinite, conic and polynomial optimization.* International Series in Operations Research & Management Science, 166. New York: Springer; 2012.

5. Arnold VI, Gusein-Zade S. Varchenko AN. *Singularities of differentiable maps. Vols. I and II.* New York: Birkhäuser/Springer; 1985.

6. Artin E. Über die Zerlegung definiter Funktionen in Quadrate. *Hamb. Abh.* 1927; 5: 100–115. The collected papers of Emil Artin, pp. 273–288. Reading: Addison-Wesley; 1965.

7. Auslender AA, Crouzeix JP. Global regularity theorems. *Math. Oper. Res.* 1988; 13(2): 243–253.

8. Bank B, Mandel R. *Parametric integer optimization.* Mathematical Research, 39, Berlin: Academie-Verlag; 1988.

9. Belousov EG. *Introduction to convex analysis and integer programming,* Moscow: Moscow University Publ. (in Russian); 1977.

10. Belousov EG, Klatte D. A Frank–Wolfe type theorem for convex polynomial programs. *Comput. Optim. Appl.* 2002; 22(1): 37–48.

11. Bertsekas DP (with Nedic A and Ozdaglar AE). *Convex analysis and optimization.* Belmont, MA: Athena Scientific; 2003.

12. Blekherman G, Parrilo PA, Thomas RR. (eds.) *Semidefinite optimization and convex algebraic geometry.* MOS–SIAM Series Optimization, 13. SIAM; 2013.

13. Bochnak J, Lojasiewicz S. A converse of the Kuiper–Kuo theorem, in *Proceedings of Liverpool Singularities Symposium, I* (1969/70), 1971, pp. 254–261. Lecture Notes in Mathematics, Springer; 1971. 192.

14. Bochnak J, Risler JJ. Sur les exposants de Lojasiewicz. *Comment. Math. Helv.* 1975; 50(4): 493–507.

15. Bochnak J, Coste M, Roy MF. *Real algebraic geometry.* Ergebnisse der Mathematik und ihrer Grenzgebiete, 36. Berlin: Springer-Verlag; 1998

16. Bolte J, Daniilidis A, Lewis AS. Generic optimality conditions for semialgebraic convex programs. *Math. Oper. Res.* 2011; 36(1): 55–70.

17. Broughton S. Milnor number and topology of polynomial hypersurfaces. *Invent. Math.* 1988; 92(2): 217–241.

18. Brézis H, Coron JM, Nirenberg L. Free vibrations for a nonlinear wave equation and a theorem of P. Rabinowitz. *Comm. Pure Appl. Math.* 1980; 33(5): 667–684.

19. Brézis H, Coron JM, Nirenberg L. Remarks on Finding Critical Points. *Comm. Pure Appl. Math.* 1991; 44(8–9): 939–963.

20. Brieskorn E, Knörrer H. *Plane algebraic curves,* Boston: Birkhäuser; 1986.

21. Choi MD, Lam TY, Reznick B. Sums of squares of real polynomials. In: Bill Jacob, Alex Rosenberg (eds.) *K-Theory and algebraic geometry. Connections with quadratic forms and division algebras: Proc. Summer Research Institute on Quadratic Forms and Division Algebras, July 6–24, 1992, University of California, Santa Barbara.* Proc. Sympos. Pure Math., 58(2). Providence, RI: American Mathematical Society; 1995; pp. 103–126.

22. Coste M. *An introduction to semialgebraic geometry.* Dip. Mat. Univ. Pisa, Dottorato di Ricerca in Matematica, Istituti Editoriali e Poligrafici Internazionali, Pisa; 2000.

23. Curto R, Fialkow L. Truncated K-moment problems in several variables. *J. Oper. Theory* 2005; 54: 189–226.

24. Demmel J, Nie JW, Powers V. Representations of positive polynomials on noncompact semialgebraic sets via KKT ideals. *J. Pure Appl. Algebra* 2007; 209(1): 189–200.

25. Đinh ST, Hà HV, Thao NT. Lojasiewicz inequality for polynomial functions on non compact domains. *Internat. J. Math.* 2012; 23(4): 1250033 (28 pages).

26. Đinh ST, Hà HV, Phạm TS. A Frank–Wolfe type theorem and Holder-type global error bound for generic polynomial systems. Available from: http://viasm.edu.vn/wp-content/uploads/2012/11/Preprint_1227.pdf [Accessed 27th November 2012].

27. Đinh ST, Hà HV, Phạm TS, Thao NT. Global Lojasiewicz-type inequality for non-degenerate polynomial maps. *J. Math. Anal. Appl.* 2014; 410(2): 541–560.

28. Đinh ST, Hà HV, Phạm TS. A Frank–Wolfe type theorem for nondegenerate polynomial programs. *Math. Program. Ser. A* 2014; 147(1–2): 519–538.

29. Đinh ST, Hà HV, Phạm TS. Hölder-type global error bounds for nondegenerate polynomial systems. Available from: http://arxiv.org/abs/1411.0859 [Accessed 4th November 2014].

30. Đinh ST, Phạm TS. Lojasiewicz-type inequalities with explicit exponents for the largest eigenvalue function of real symmetric polynomial matrices. *Internat. J. Math.* 2016; 27(2): 1650012 (27 pages).

31. Đinh ST, Phạm TS. Łojasiewicz inequalities with explicit exponent for smallest singular value functions. Available from: http://arxiv.org/abs/ 1604.02805 [Accessed 11th April 2016].

32. Đoạt DV, Hà HV, Phạm TS. Well-posedness in unconstrained polynomial optimization problems. *SIAM J. Optim.* 2016; 26: 1411–1428.

33. Dontchev AL, Zolezzi T. *Well-Posed optimization problems.* Lecture Notes in Mathematics, 1543. Berlin: Springer-Verlag; 1993.

34. van den Dries L. O-minimal structures and real analytic geometry. In: Mazur B, Schmid W, Yau ST, Jerison D, Singer I, Stroock D. (eds.) *Current developments in mathematics, Proc. Seminar held in Cambridge*, MA, 1998. Somerville, MA: Int. Press; 1999. pp. 105–152.

35. Drusvyatskiy D, Ioffe AD, Lewis AS. Generic minimizing behavior in semi-algebraic optimization. *SIAM J. Optim.* 2016; 26(1): 513–534

36. Durfee A. The index of $\operatorname{grad} f(x, y)$. *Topology* 1998; 37(6): 1339–1361.

37. Ekeland I. On the variational principle. *J. Math. Anal. Appl.* 1974; 47: 324–353.

38. Forti M., Tesi A. The Łojasiewicz exponent at an equilibrium point of a standard CNN is 1/2. *Internat. J. Bifur. Chaos Appl. Sci. Engrg.* 2006; 16(8): 2191–2205.

39. Frank M, Wolfe P. An algorithm for quadratic programming. *Naval Res. Logist. Quart.* 1956; 3: 95–110.

40. Fujiwara O. Morse programs: a topological approach to smooth constrained optimization. I. *Math. Oper. Res.* 1982; 7(4): 602–616.

41. Fujiwara O. A note on differentiability of global optimal values. *Math. Oper. Res.* 1985; 10(4): 612–618.

42. Garey MR, Johnson DS. *Computers and intractability: A guide to the theory of NP-completeness.* San Francisco, CA.: W.H. Freeman & Company; 1979.

43. Gindikin SG. Energy estimates connected with the Newton polyhedron. (Russian) *Trudy Moskovskogo Matematicheskogo Obshchestva.* 1974; 31: 189–236. (English) *Trans. Moscow Math. Soc.* 1974; 31: 193–246.

44. Goresky M, MacPherson R. *Stratified Morse theory.* Berlin: Springer-Verlag; 1988.

45. Guillemin V, Pollack A. *Differential topology,* Reprint of the 1974 original. Providence, RI: AMS Chelsea Publishing; 2010.

46. Gwoździewicz J. The Łojasiewicz exponent of an analytic function at an isolated zero. *Comment. Math. Helv.* 1999; 74(3): 364–375.

47. Hà HV, Phạm TS. Minimizing polynomial functions. *Acta Math. Vietnam* 2007; 32(1): 71–82.

48. Hà HV, Phạm TS. An estimation of the number of bifurcation values for real polynomials. *Acta Math. Vietnam* 2007; 32(2–3): 141–153.

49. Hà HV, Phạm TS. Critical values of singularities at infinity of complex polynomials. *Vietnam J. Math.* 2008; 36(1): 1–38.

50. Hà HV, Phạm TS. On the Łojasiewicz exponent at infinity of real polynomials. *Ann. Polon. Math.* 2008; 94(3): 197–208.

51. Hà HV, Phạm TS. Global optimization of polynomials using the truncated tangency variety and sums of squares,. *SIAM J. Optim.* 2008; 19(2): 941–951.

52. Hà HV, Phạm TS. Solving polynomial optimization problems via the truncated tangency variety and sums of squares. *J. Pure Appl. Algebra* 2009; 213(11): 2167–2176.

53. Hà HV, Phạm TS. Representations of positive polynomials and optimization on noncompact semialgebraic sets. *SIAM J. Optim.* 2010; 20(6): 3082–3103.

54. Hà HV. Global Hölderian error bound for non-degenerate polynomials. *SIAM J. Optim.* 2013; 23(2): 917–933.

55. Hadamard J. Sur les problèmes aux dérivees partielles et leur signification physique. *Bull. Univ. Princeton* 1902; 13: 49–52.

56. Hanzon B, Jibetean D. Global minimization of a multivariate polynomial using matrix methods. *J. Global Optim.* 2003; 27(1): 1–23.

57. Hiriart-Urruty JB, A short proof of the variational principle for approximate solutions of a minimization problem. *Amer. Math. Monthly* 1983; 90(3): 206–207.

58. Hoffman AJ. On approximate solutions of linear inequalities. *J. Research Nat. Bur. Standards* 1952; 49: 263–265.

59. Ioffe AD, Lucchetti RE, Revalski JP. Almost every convex or quadratic programming problem is well posed. *Math. Oper. Res.* 2004; 29(2): 369–382.

60. Ioffe AD, Lucchetti RE. Generic well-posedness in minimization problems. *Abstr. Appl. Anal.* 2005; 4: 343–360.

61. Ioffe AD. An invitation to tame optimization. *SIAM J. Optim.* 2008; 19(4): 1894–1917.

62. Jelonek Z. Geometry of real polynomial mappings. *Math. Z.* 2002; 239(2): 321–333.

63. Jeyakumar V., Phạm TS., Li G. Convergence of the Lasserre hierarchy of SDP relaxations for convex polynomial programs without compactness. *Oper. Res. Lett.* 2014; 42(1): 34–40.

64. Jibetean D, Laurent M. Semidefinite approximations for global unconstrained polynomial optimization. *SIAM J. Optim.* 2005; 16(2): 490–514.

65. Jongen H Th., Jonker P., Twilt F. *Nonlinear optimization in finite dimensions.* Nonconvex Optim. Appl., 47. Dordrecht: Kluwer Academic Publishers; 2000.

66. Khovanskii AG. Newton polyhedra and toroidal varieties. *Funct. Anal. Appl.* 1978; 11: 289–296.

67. Klatte D. Hoffman's error bound for systems of convex inequalities. in *Mathematical programming with data pertubations.* Lecture Notes in Pure and Appl. Math., 195, New York: Dekker; 1998. pp. 185–199.

68. Klatte D, Li A. Asymptotic constraint qualifications and global error bounds for convex inequalities. *Math. Program.* 1999; 84(1): 137–160.

69. de Klerk E, Laurent M. On the Lasserre hierarchy of semidefinite programming relaxations of convex polynomial optimization problems. *SIAM J. Optim.* 2011; 21(3): 824–832.

70. Kollár J. An effective Łojasiewicz inequality for real polynomials. *Period. Math. Hungar.* 1999; 38(3): 213–221.

71. Kouchnirenko AG. Polyhèdres de Newton et nombre de Milnor. *Invent. Math.* 1976; 32(1): 1–31.

72. Krivine JL. Anneaux préordonnés. *J. Anal. Math.* 1964; 12: 307–326.
73. Kurdyka K, Orro P, Simon S. Semialgebraic Sard theorem for generalized critical values. *J. Differential Geom.* 2000; 56(1): 67–92.
74. Kurdyka K, Spodzieja S. Separation of real algebraic sets and the Łojasiewicz exponent. *Proc. Amer. Math. Soc.* 2014; 142(9): 3089–3102.
75. Lasserre JB. Optimisation globale et théorie des moments. *C. R. Acad. Sci. Paris Sér. I Math.* 2000; 331(11): 929–934.
76. Lasserre JB. Global optimization with polynomials and the problem of moments. *SIAM J. Optim.* 2001; 11(3): 796–817.
77. Lasserre JB. An explicit exact SDP relaxation for nonlinear 0–1 programs. In: Aardal K, Gerards B. (eds.) *Proc. 8th Integer Programming and Combinatorial Optimization Conf. (IPCO).* Lecture Notes in Comput. Sci., 2081; Berlin: Springer; 2001. pp. 293–303.
78. Lasserre JB. Convexity in semialgebraic geometry and polynomial optimization. *SIAM J. Optim.* 2008; 19(4): 1995–2014.
79. Lasserre JB. *Moments, positive polynomials and their applications.* London: Imperial College Press; 2009.
80. Lasserre JB. *Introduction to polynomial and semi-algebraic optimization.* Cambridge: Cambridge University Press; 2015.
81. Laurent M. Semidefinite representations for finite varieties. *Math. Program. Ser. A.* 2007; 109(1): 1–26.
82. Laurent M. Sums of squares, moment matrices and optimization over polynomials. in Putinar M, Sullivant S. (eds.) *Emerging applications of algebraic geometry.* IMA Vol. Math. Appl. 149. New York: Springer; 2009. pp. 157–270.
83. Laurent M. Optimization over polynomials: Selected topics. in Jang SY, Kim YR, Lee DW, Yie I. (eds.) *Proc. Int. Congr. Mathematicians.* Seoul: Kyung Moon SA Co. Ltd.; 2014. pp. 843–869.
84. Lee GM, Phạm TS. Stability and genericity for semialgebraic compact programs. *J. Optim. Theory Appl.* 2016; 169(2): 473–495.
85. Lee GM, Phạm TS. Generic properties for semialgebraic programs. Available from: http://www.optimization-online.org/DB_HTML/2015/06/4957.html [Accessed 12th June 2015].
86. Lewis AS, Pang JS. Error bounds for convex inequality systems. in Crouzeix JP, Martinez-Legaz JE, Volle M. (eds) *Generalized convexity, generalized monotonicity: recent results.* Nonconvex Optim. Appl., 27. Dordrecht: Kluwer Academic Publishers; 1998. pp. 75–110.
87. Lewis AS. Nonsmooth optimization: conditioning, convergence and semialgebraic models. in Jang SY, Kim YR, Lee DW, Yie I. (eds.) *Proc. Int. Congre. Mathematicians.* Seoul: Kyung Moon SA Co. Ltd.; 2014. pp. 871–895.
88. Li G. On the asymptotic well behaved functions and global error bound for convex polynomials. *SIAM J. Optim.* 2010; 20(4): 1923–1943.
89. Li G. Global error bounds for piecewise convex polynomials. *Math. Program. Ser. A* 2013; 137(1–2): 37–64.
90. Li G, Mordukhovich BS, Phạm TS. New fractional error bounds for polynomial systems with applications to Hölderian stability in optimization and spectral theory of tensors. *Math. Program. Ser. A* 2015; 153(2): 333–362.

91. Li G, Mordukhovich BS, Nghia TTA, Phạm TS. Error bounds for parametric polynomial systems with applications to higher-order stability analysis and convergence rates. *Math. Program. Ser. B* 2016; DOI 10.1007/s10107-016-1014-6.

92. Loi TL, Zaharia A. Bifurcation sets of functions definable in *o*-minimal structures. *Illinois J. Math.* 1998; 42(3): 449–457.

93. Łojasiewicz S. Ensembles semi-analytiques. Preprint, l'Institut des Hautes Études Scientifiques, Bures-sur-Yvette, 1965.

94. Luo ZQ, Pang JS. Error bounds for analytic systems and their applications. *Math. Program. Ser. A* 1994; 67(1): 1–28.

95. Luo ZQ, Zhang S. On extensions of the Frank–Wolfe theorems. *Comput. Optim. Appl.* 1999; 13(1–3): 87–110.

96. Luo ZQ, Sturm JF. Error bound for quadratic systems. In: Frenk H, Roos K, Terlaky T, Zhang S. (eds.) *High performance optimization.* Appl. Optim., 33. Dordrecht: Kluwer Academic Publishers; 2000. pp. 383–404.

97. Mangasarian OL. A condition number for linear inequalities and linear programs. in Bamberg G, Opitz O. (eds.) *Proc. 6th Sympos. über Operations Research,* Augsburg (BRD), 7–9 September 1981, Methods of Operations Research, Part 1, Vol. 43, Verlagsgruppe Athenaum/Hain/Scriptor/Hanstein (Konigstein 1981), pp. 3–15.

98. Marshall M. Representation of non-negative polynomials having finitely many zeros. *Ann. Faculte Sci. Toulouse* 2006; 15(6), 599–609

99. Marshall M. *Positive polynomials and sums of squares.* Math. Surveys and Monographs, 146. Providence, RI: American Mathematical Society; 2008.

100. Mikhalov VP. The behaviour at infinity of a class of polynomials. in (Russian) *Trudy Mat. Inst. Steklov.* 1967; 91: 59–80.

101. Milnor J. *Singular points of complex hypersurfaces.* Annals of Mathematics Studies, 61. New York: Princeton University Press; 1968.

102. Milnor J. *Morse theory.* 5th ed., Annals of Mathematics Studies, 51. New York: Princeton University Press; 1963.

103. Mordukhovich BS. *Variational analysis and generalized differentiation, I: Basic Theory; II: Applications.* Berlin: Springer-Verlag; 2006.

104. Morse M. Relations between the critical points of a real function of n independent variables. *Trans. Amer. Math. Soc.* 1925; 27(3): 345–396.

105. Motzkin T. The arithmetic-geometric inequalities, in Shisha O. (ed.) *Inequalities. Proc. Sympos. Wright-Patterson Air Force Base,* Ohio, August 19–27, 1965, Academic Press; 1967, pp. 205–224.

106. Némethi A, Zaharia A. Milnor fibration at infinity. *Indag. Math.* 1992; 3(3): 323–335.

107. Nesterov Y. Squared functional systems and optimization problems. In: Frenk H, Roos K, Terlaky T, and Zhang S. (eds.) *High performance optimization.* Appl. Optim., 33. Dordrecht: Kluwer Academic Publishers; 2000. pp. 405–440.

108. Ng KF, Zheng XY. Global error bounds with fractional exponents. *Math. Program. Ser. B* 2000; 88(2): 357–370.

109. Ngai HV, Thera M. Error bounds for systems of lower semicontinuous functions in Asplund spaces. *Math. Program. Ser. B* 2009; 116(1–2): 397–427.

110. Nie JW, Demmel J, Sturmfels B. Minimizing polynomials via sum of squares over the gradient ideal. *Math. Program. Ser. A* 2006; 106(3): 587–606.

111. Nie JW. An exact Jacobian SDP relaxation for polynomial optimization. *Math. Program. Ser. A* 2013; 137(1–2): 225–255.

112. Nie JW. Polynomial optimization with real varieties. *SIAM J. Optim.* 2013; 23(3): 1634–1646.

113. Nie JW. Certifying convergence of Lasserre's hierarchy via flat truncation. *Math. Program. Ser. A* 2013; 142(1): 485–510.

114. Nie JW. Optimality conditions and finite convergence of Lasserre's hierarchy. *Math. Program. Ser. A.* 2014; 146(1–2): 97–121.

115. Obuchowska WT. On generalizations of the Frank–Wolfe theorem to convex and quasi-convex programmes. *Comput. Optim. Appl.* 2006; 33(2–3): 349–364.

116. Oka M. On the bifurcation of the multiplicity and topology of the Newton boundary. *J. Math. Soc. Japan* 1979; 31(3): 435–450.

117. Oka M. *Non-degenerate complete intersection singularity.* Actualités Mathématiques. Paris: Actualités Mathématiques Hermann; 1997.

118. Palais RS, Smale S. A generalized Morse theory. *Bull. Amer. Math. Soc.* 1964; 70: 165–172.

119. Parrilo PA. *Structured semidefinite programs and semialgebraic geometry methods in robustness and optimization.* Ph.D. thesis, California Institute of Technology, May 2000. Available at resolver.caltech.edu/CaltechETD: etd-05062004-055516.

120. Parrilo PA. *An explicit construction of distinguished representations of polynomials nonnegative over finite sets.* IfA Technical Report AUT02-02, 002; 2002. Available from: http://www.mit.edu/~parrilo, ETH Zürich.

121. Parrilo PA, Sturmfels B. Minimizing polynomial functions. In: Basu S, Gonzalez-Vega L. (eds.) *Algorithmic and quantitative real algebraic geometry.* DIMACS Ser. Discrete Math. Theoret. Comput. Sci., 60. Providence, RI: American Mathematicla Society; 2003. pp. 83–99.

122. Parrilo PA. Semidefinite programming relaxations for semialgebraic problems. *Math. Program. Ser. B* 2003; 96(2): 293–320.

123. Pang JS. Error bounds in mathematical programming. *Math. Program. Ser. B* 1997; 79(1–3): 299–332.

124. Pataki G, Tunçel L. On the generic properties of convex optimization problems in conic form. *Math. Program. Ser. A* 2001; 89(3): 449–457.

125. Perold AF. Generalization of the Frank–Wolfe theorem. *Math. Program.* 1980; 18(2): 215–227.

126. Phạm TS. An explicit bound for the Łojasiewicz exponent of real polynomials. *Kodai Math. J.* 2012; 35(2): 311–319.

127. Phạm TS, Truonng XD, Yao JC. Weak sharp minima for semi-algebraic vector optimization problems. Available from: http://www.optimization-online.org/DB_HTML/2014/10/4592.html [Accessed 10th October 2014].

128. Powers V, Wörmann T. An algorithm for sums of squares of real polynomials. *J. Pure Appl. Algebra* 1998; 127(1): 99–104.

129. Prestel A, Delzell CN. *Positive polynomials. From Hilbert's 17th problem to real algebra.* Berlin: Springer-Verlag; 2001.

130. Putinar M. Positive polynomials on compact semi-algebraic sets. *Indiana Univ. Math. J.* 1993; 42(3): 969–984.

131. Rabier PJ. Ehresmann fibrations and Palais-Smale conditions for morphisms of Finsler manifolds. *Ann. of Math.* 1997; 146(3): 647–691.

132. Reznick B. Some concrete aspects of Hilbert's 17th problem. in N. Delzell CN., Madden JJ. (eds.) *Real algebraic geometry and ordered structures.* Contemp. Math. 253. Providence, RI: American Mathematical Society; 2000. pp. 251–272.

133. Robinson S. Regularity and stability of convex multivalued functions. *Math. Oper. Res.* 1976; 1(2): 130–143.

134. Rockafellar RT, Wets R. *Variational analysis.* Grundlehren Math. Wiss., 317. Berlin: Springer-Verlag; 1998.

135. Saigal R, Simon C. Generic properties of the complementarity problem. *Math. Program.* 1973; 4: 324–335.

136. Scheiderer C. Positivity and sums of squares: A guide to recent results. in Putinar M, Sullivant S. (eds.) *Emerging applications of algebraic geometry.* IMA Vol. Math. Appl., 149. New York: Springer; 2009. pp. 271–324.

137. Schmüdgen K. The K-moment problem for compact semi-algebraic sets. *Math. Ann.* 1991; 289(2): 203–206.

138. Schweighofer M. Optimization of polynomials on compact semialgebraic sets. *SIAM J. Optim.* 2005; 15(3): 805–825.

139. Schweighofer M. Global optimization of polynomials using gradient tentacles and sums of squares. *SIAM J. Optim.* 2006; 17(3): 920–942.

140. Shapiro A. First and second order analysis of nonlinear semidefinite programs. *Math. Program. Ser. B* 1997; 77(2): 301–320.

141. Shor NZ. An approach to obtaining global extremums in polynomial mathematical programming problems. *Kibernetika* 1987; 5: 102–106.

142. Spingarn JE, Rockafellar RT. The generic nature of optimality conditions in nonlinear programming. *Math. Oper. Res.* 1979; 4(4): 425–430.

143. Spingarn JE. On optimality conditions for structured families of nonlinear programming problems. *Math. Program.* 1982; 22(1): 82–92.

144. Stengle G. A Nullstellensätz and a Positivstellensätz in semialgebraic geometry. *Math. Ann.* 1974; 207: 87–97.

145. Tykhonov AN. On the stability of the functional optimization problem. *USSR J. Compo Math. Math. Phys.* 1966; 6: 631–634.

146. Vandenberghe L, Boyd S. Semidefinite programming. *SIAM Rev.* 1996; 38(1): 49–95.

147. Yang WH. Error bounds for convex polynomials. *SIAM J. Optim.* 2008; 19(4): 1633–1647.

Index

Printed in the United States
By Bookmasters